Damiano Anselmi

From Physics to Life

A journey to the infinitesimally small and back

Copyright © 2019 Damiano Anselmi

All rights reserved.

ISBN: 9781089441144

Contents

Physics and Special Effects 9

From Macro to Micro: Quantum Mechanics 15

Life is Freedom. And Freedom is Life 35

What are We? Natural Life and Artificial Life 43

Our Future, our Destiny 55

From the Small to the Infinitesimally Small: Quantum Gravity . 65

At the Heart of the Universe, where Everything Loses its Significance, yet still Governs the Cosmos 77

Discovering the Hidden Meaning of the Universe 93

The Correspondence between the Large and the Small 107

Our Limits 117

The Threat of Involution 127

The discoveries in physics over the past century provide answers to the deepest questions we can ask (who are we, where do we come from, what is the meaning of life, what is the origin of the universe, what is the meaning of it all?). By presenting the key principles of quantum mechanics in a new light and combining them with very recent ideas about quantum gravity, we can understand what nature is trying to tell us about life and the meaning of the universe.

1
Physics and Special Effects

Physics is often used by science communicators to amaze, captivate, and fascinate the reader, employing a wide array of special effects. How many times have you heard about the mysterious black hole, where everything can fall in but nothing can escape, to the extent that even light cannot emerge from it? Not to mention the Big Bang and the immensity of the universe, astronomical distances, the diversity of galaxies, which make us ponder our solitude in the cosmos and give us a sense of smallness and powerlessness. And, of course, the mysteries associated with the concept of time, because when we observe a distant galaxy, we are seeing it in the past since the light reaching us today necessarily departed many years ago. These are very interesting subjects. However, compared to what we will discuss here, they are hardly more than curiosities, because what we will deal with is not curious, fascinating, or unexpected: it is unsettling. That's why we must prepare ourselves mentally.

Space, time: we thought they were absolute, but they are actually relative. We thought they were separate, but they are, in reality, connected in the notion of space-time. We thought we had understood, with a certain degree of precision, the reality around us, or a part of it, but then we discovered that things were different from what we had imagined. In many cases, these advances, or updates, in our knowledge are very interesting, but they are not earth-shattering. Many popular science books deal with the captivating aspects of discoveries in physics but carefully avoid, one might say, delving into what can truly unsettle us, what can forever change the way we think about ourselves and the world.

In this book, I won't seek to amaze you with special effects. Instead, I'll simply disrupt your thinking by addressing a series of recurring questions. Who are we? Where do we come from? How are we made? Where are we going? Do we have a role in the universe? A mission, perhaps? What is the reality that surrounds us? What is being? What is the universe itself? What can physics tell us about these matters? I assume that many readers have been eagerly awaiting answers to these questions for a long time. It's time to reveal them.

Many of us probably watched science popularization shows on television when we were children. In many countries around the world, such programs deal with fascinating subjects, especially when it comes to discoveries in physics. However, I bet that when they attempt to explain the basics of quantum mechanics to viewers, it's not that easy to understand what they are saying. At first glance, quantum mechanics doesn't seem like a topic that leaves much room for imagination. Perhaps it's not as captivating for the reader, or viewer, as topics like cosmic subjects, galaxies, relativity, or the Big Bang. In fact, for many, it might appear to be a purely technical and dry, even boring, subject. The truth is quite the opposite, and I'm sure that after finishing reading this book, you will agree with me.

Science communicators face a challenging task when it comes to quantum mechanics. Generally, they prefer to focus on important, but peripheral aspects, rather than the fundamental concepts and the most problematic ideas.

There are several reasons for this discomfort with quantum mechanics. The most important one is that physicists themselves have been, to say the least, hesitant about it. For a very long time, many of them refused to accept its message, no matter how disruptive it was, and to develop it further. In turn, this slowed down the popularization process and reduced the efforts that could have allowed the general public to appreciate fundamental concepts in a simple manner.

Some physicists, including Einstein, never accepted the radical and complete revolution brought about by quantum mechanics. They were unwilling to consider it a definitive theory, precisely because its message was unacceptable to them. Indeed, it shattered all the certainties they had taken for granted until then. Often, humans prefer to continue on the path of their beliefs, the certainties they hold and have become "attached" to, instead of questioning all of that to better understand what is right there, in front of their eyes.

In the end, here is the key to every understanding: the ability to see what is right in front of our eyes.

In this book, we will embark on a "vertical" journey. We will delve into the depths of the universe to understand who we are, where we come from, and what relationship exists between us and the rest of reality. We will even come to understand our purpose in the universe, the reason for our existence. Indeed, over the past century, physics has made tremendous progress and is capable of answering most of our questions. However, in recent times it has undergone a major crisis, like perhaps all of science. We will see why that is as well.

We will divide this journey into two parts. First, we will descend from our scales of magnitude, which we can quantify in meters or centimeters, to atomic scales, where quantum mechanics reigns. In about a centimeter, approximately one hundred million atoms can be aligned. So, we will make our first and only stop at one-hundredth of a millimeter. Do these distances seem small to you? They might have been at the beginning of the 20th century, but today, they are not as small. We now have some familiarity with the phenomena that occur at this scale of magnitude. We will linger there for a while because the implications of what we will find are earth-shattering.

In the second part of the journey, we will descend from atomic scales to scales that are a billion billion times smaller, where we will encounter quantum gravity. It may seem astonishing that we can make such a significant leap all at once, but that's precisely what we will manage to do, for reasons we will analyze in detail. We will learn many new things, particularly concerning what makes sense and what does not, what has fundamental meaning in nature, and what can have at most an approximate meaning, useful for daily life, but limited and misleading whenever we want to ask fundamental questions about ourselves and nature.

But most importantly, the journey we are about to undertake is a round trip. In fact, the return will be the most crucial part, as it will reveal to us who we are, where we come from, and where we are going.

Indeed, we can understand many things about ourselves and the universe if we try to grasp nature as a whole, using the knowledge accumulated by science throughout history. When we move "horizontally," for example, from one planet to another, we do not find significant changes, setting aside, for a moment, the fact that Earth has life while other planets do not (we will later understand the meaning of this difference). Traveling from Mars to Venus, from Jupiter to Saturn, we encounter gas or solid planets, with or without atmospheres, filled with acids or not. We can go from medium-sized stars like our Sun to smaller or larger stars, eventually reaching supernovae, white dwarfs, and even black holes. We can even imagine traveling from one galaxy to another. Each will have its own shape and history. We would encounter a wide variety of options, configurations, and different situations due to diverse circumstances. This is a type of variety we might well call "horizontal." In other words, very interesting, perhaps fascinating, but not earth-shattering.

On the contrary, when we embark on a vertical journey, descending through relative scales of magnitude, we encounter truly unsettling changes. But there's no need to worry too much about it. We will come to know them until they become more familiar to us than we can imagine. To the point that once we draw the natural conclusions they lead us to, we will have to ask ourselves, "How could we not have thought of this before?" What our journey will teach us has always been right before our eyes, and we simply refused to see it.

This type of journey necessarily requires diving deep rather than ascending to cosmic scales of magnitude. On one hand, as mentioned earlier, what we find when ascending to astronomical distances is interesting but not earth-shattering. On the other hand, we are made up of what is smaller than us, not what is larger than us. Therefore, we can learn much more about ourselves by descending into the abyss of the infinitely small, rather than getting lost in the vast expanses of cosmic space.

2

From Macro to Micro: Quantum Mechanics

My scientific attitude has always been that in the face of nature, we must put ourselves in the shoes of a child, a primary school student. Nature is the teacher, and we must simply listen to her and strive to understand what she is trying to tell us. We are not the professors; we are not the ones who can dictate to reality what it should tell us. This might seem like an obvious, mundane consideration, but it's important to note that, especially in modern times, it's not the position shared by most physicists. Often, invoking constraints and artificial constructs, deliberately or unintentionally, many of them insist on seeing in reality what reality repeatedly refuses to tell them, even after numerous failures.

When nature seems to be heading in a direction contrary to our expectations, our first reaction is not to change our ideas to align with nature's message. The initial response is to artificially complicate matters to try to fit the novelty, or even a domesticated version of it, into our preconceived framework of ideas. Unfortunately, it is the most natural human reaction, even for many physicists, who, when exploring nature, try to find the message that suits their preferences, rather than adopting the perspective of a child, a primary school student, who says to the teacher, i.e., nature: "I am a blank slate, ready to learn what you have to teach me."

Quantum mechanics conveys a profoundly unsettling message that we are called to decipher without prejudice. Combined with other discoveries in physics over the past century, including the most recent ones to which I have had the honor of contributing, it allows us to paint a new picture of the universe.

To appreciate all of this, we compare the phenomena that occur at our scale of magnitude, that is to say, relative distances measuring around meters or centimeters, to phenomena occurring at much smaller scales. We start with atomic distances, like the radius of an atom, which is a hundred million times smaller than a centimeter. Later, we will push on to "infinitesimally" small distances, equivalent to one billionth of a billionth of a billionth of a centimeter, where the effects, direct or indirect, of quantum gravity begin to manifest themselves. As one can easily imagine, the reality of things changes completely when we dive into such depths. We, as human beings, are placed in an environment where certain types of phenomena occur, codified by certain physical laws. Well, the phenomena occurring in the microscopic world are so profoundly different from what we are familiar with that we might find ourselves entirely incapable of describing them using the words of our language.

We cannot take for granted that nature is meant to be explained or understood by us. Nature is what it is. It is indifferent to us. We, on the other hand, are one of the many living species on one of the many planets. From a physical perspective, a living being is an arrangement of atoms, just one of the many ways atoms come together. "Our" arrangement of atoms is attempting to comprehend what happens at scales equivalent to those of the smallest ingredient it is composed of, the atom itself, and then at scales a billion billion times smaller. It's trying to understand what goes on down there to gain additional insight into the nature of the world and the meaning of what happens up here. Most likely, this endeavor could lead to a short circuit, a vicious circle, or a problem that is entirely unsolvable.

What we learn from the world around us is absolutely partial and insufficient because the world itself that surrounds us is just a part: it's a small piece of the universe. At the same time, it shapes our thinking and our way of seeing things. In fact, very often, the intuition suggested by the world around us misleads us because the reality down there is completely different from the reality up here. Today, we can precisely describe this difference, which I am about to do in this book.

We are finally able to understand how profound the difference between the two worlds, the realm of what is large and the realm of what is small, truly is. The information we are about to learn is so bewildering that it will make us doubt whether nature has endowed us with sufficient abilities to overcome the obstacles we encounter on the path to understanding nature itself. It makes us fear that at some point, we will be compelled to surrender.

Just to give you a glimpse of what awaits us, consider that, in the world around us, we can perceive, we can "see" what is happening, simply because we can "turn on the light." What does this mean? Light is a stream of billions and billions of photons, the quanta of light, which are the "corpuscles" or "elementary particles" of light, its fundamental and indivisible units. Photons can originate from any source, such as the sun or a light bulb, for example. They strike the objects around us. Some of them are absorbed, and some are reflected. Many of the reflected photons from objects enter our eyes and generate nerve signals, which then reach our brain and are processed to produce what we call images. The quanta of light that fill the space around us are incredibly numerous. Thus, we have the perception of seeing a continuous reality before our eyes.

Thanks to this, we have a clear perception of all the things around us. At macroscopic scales, I can see this table in front of me, and I can say that it is there and stationary. I see a car racing by at high speed, and I have a clear idea of its position at the moment I see it. I also have a clear idea of its speed, which I can potentially measure precisely using a speed camera. We can perceive and measure both the position and velocity of all the objects that surround us.

It is of fundamental importance to be able to perceive both because it is the combination of the two that gives us the perception of motion, and therefore, of temporal continuity. At our scales of magnitude, we can track the trajectory of an object with the precision we desire. If we accurately perceived only the position but not the velocity, we would not be able to connect an earlier event to a later event. It would be somewhat like looking at a series of photographs, which can be as precise as you like but remain static. If we compare watching photos to watching a video, we easily realize what we lose.

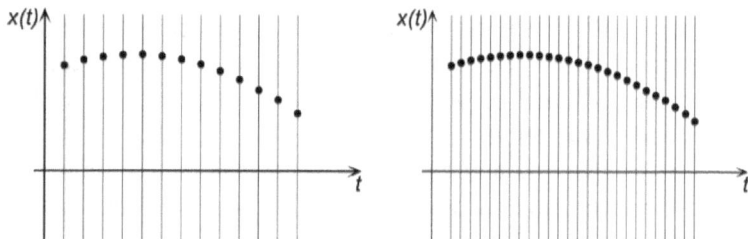

Fig. 1 | Temporal continuity allows us to identify the trajectories of objects

However, down there, at small distances, total darkness prevails. This already happens at the atomic scale, not to mention a billion billion times below that. What does "total darkness" mean? It means there is no possibility of "turning on the light." Indeed, if we wanted, for example, to "turn on the light" to "see" an atom, we would have to consider that light is made of photons. In practice, we would end up inundating the atom with a tidal wave of photons, completely and instantly disrupting its state. In the end, we would not see reality as it is but as we have altered it.

Is there a possibility of seeing reality as it is at such distances? Could we, for example, send a photon to the atom one by one to disturb it as little as possible? No, because there are two possibilities: either the photon we send to the atom is too "large" compared to the atom, and then it does not allow us to measure its position precisely, or it is too "energetic," and then it does not allow us to measure its velocity. If we want to measure both at the same time, the only possibility we have is to inundate the atom with a photon tsunami, which only works with macroscopic objects.

More precisely, if we send a low-energy photon to the atom to disturb it as little as possible, we cannot locate the atom because a low-energy photon is "too large" to precisely measure its position. In other words, it has a long wavelength. In a sense, it is a "big and bulky" photon. It would be like trying to determine the location of a grain of sand by throwing a one-ton boulder at it and (in the total darkness where we are forced to work) deducing the position of the grain of sand from the result of the collision with the boulder. It is evident that we cannot obtain a satisfactory result in this manner.

If we really want to determine the position of the atom, we can use a photon with a very small wavelength. In other words, a "thin" photon, i.e., smaller than the atom itself. However, in this case, the photon would be so "energetic," i.e., it would have so much energy, that it would eject the atom to very large distances. This would prevent us from determining its velocity. When we conduct an experiment of this kind, we can locate the atom, but we cannot say whether it is stationary or in motion. In reality, we can only determine where the atom *was before* being disturbed and ejected, but we cannot determine what its velocity *was*.

We are beginning to grasp the fact that the microscopic world has properties so different from the world around us that we must make significant concessions, concessions capable of disrupting, as we will see, our own understanding of reality. We can determine the position of the atom with the precision we desire, but then we must completely give up saying how fast it is moving. Or we can determine its speed by giving up determining its position. Or we can settle for an average, but then we must measure both position and velocity imprecisely, and the product of these inaccuracies has a minimum limit, so it cannot be reduced at will. Goodbye to the possibility of making a video of microscopic reality!

We have explained what is called the uncertainty principle. It sets a limit beyond which our observation cannot go. Beyond this limit, it makes no sense to talk about "determining" reality.

We must come to terms with the fact that at the microscopic level, there are questions that do not even make sense. The simple fact that we cannot simultaneously perceive, with sufficient precision, both position and velocity, destroys any notion of temporal continuity. Perception is thus reduced to a succession of flashes. The subsequent images are too different from the previous ones, or separated by too long time intervals, to allow us to reconstruct a "video." We must resign ourselves to it: it is impossible to create a video of microscopic reality. It makes no sense to claim otherwise. So, how does microscopic reality "appear"? What is it?

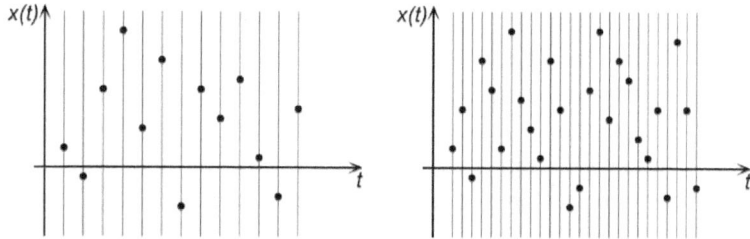

Fig. 2 | We cannot observe the "trajectory" of an atom

To provide a clearer idea, let's assume we observe the position $x(t)$ of an object at successive moments $t_1, t_2, ... t_n$. In Figure 1, we have seen what we obtain with a macroscopic object. In this case, it is easy to identify the underlying trajectory. But the situation changes radically when we observe an atom, as shown in Figure 2. At the beginning (left image), we measure the position at time intervals, let's say, one second apart. To our eyes, no clear trajectory appears. Let's stay calm, let's not get discouraged. Let's decide to space out the measurements so that only half a second passes between the previous and the next measurement (right image). Unfortunately, even in this case, we are not able to discern even a hint of a continuous trajectory. We can continue indefinitely, spacing out the measurements as little as we want, separating them by a tenth of a second, then a hundredth of a second, then a thousandth, and so on, but the situation will not change: the atom will continue to jump from one position to another without continuity, without a pattern. It seems unwilling to "stay in line."

The most we can do is settle for a very coarse and blurry trajectory, giving up the precision of measurements, so that every time we reveal an approximate position, we displace the atom as little as possible. We wanted to know what reality looks like on the microscopic scale? This is just a glimpse...

The experiments we conduct in our daily lives give us an idea of the world. They suggest physical laws and even, perhaps, "first principles." And we often tend to believe that the understanding of the world that emerges from observing a small part of it, the macroscopic world around us, is universal. But could it be a blunder, instead? A rough approximation, emerging from something smaller, constructed in an entirely different way? Are we really allowed to think that the macroscopic world has such a precise relationship with the first principles of the universe that it allows us to grasp them so easily from here, without even delving deep to verify? Are we not simplifying things a bit too much every time we think that? Today, thanks to quantum mechanics, we can say with certainty that these doubts are well-founded.

When we explore the infinitely small, we face a multitude of difficulties. We must work in total darkness. It is impossible to turn on the light. Nothing can be observed without disturbing it. We have an atom here, a photon passing by there, and all around, nothingness, absolute silence. We must deal with discrete, sporadic quantities, flashes, far from the continuum of daily reality. So how can we connect the flashes? How can we understand the laws they obey? We cannot make things too simple because they are not simple. And when we descend to even smaller distances, even deeper disruptions emerge, as you can imagine. At some point, we may be forced to make sacrifices, including the one of surrendering ourselves. Then nature will say to us, "Dear human, you have reached the end of your journey in this world. You have done your part, it was good, but now you can have a rest. Make way for somebody else!"

To better describe the significant difference between the two worlds, the microscopic and the macroscopic, we must add that when we try to describe the former, we struggle to use the words of our language correctly. We have developed a language, and therefore a way of thinking, by interacting with the macroscopic world around us, which, as mentioned earlier, gives us a very limited and partial perception of reality. When we explore the microscopic world, we discover that the same words we commonly use are inadequate, so inadequate that we even have difficulty using the verb "to be," which is the cornerstone of any language. As we ponder what "is," what the universe is, it becomes clear that if the verb "to be" were to disappear, all the questions we ask would lose their meaning completely.

Why can't we use the verb "to be" with ease anymore? Quantum mechanics explains it to us. When we descend to small distances, into the depths of the darkest obscurity, we can observe our atom here and there, now and then, but we cannot observe it continuously because observing essentially means throwing something of comparable size at it, shattering it, or trapping it with a detector. In other words, as mentioned earlier, it often means disturbing or even completely distorting its state. On the macroscopic scale, we can conduct experiments without these problems, that is, observe the objects around us without basically disturbing them, because the light that strikes them, in general, doesn't move them, heat them up much, or significantly change their state. But on the microscopic scale, this is not possible. And we certainly cannot build instruments smaller than an atom, because instruments are made up of atoms. That's why we have to argue indirectly. But in this way, it's evident that we encounter a huge problem, namely that at the very moment we want to observe reality, we disturb it to the point of destroying what we wanted to observe. So, what does "observing reality" really mean? If it's no longer clear what "observing" means, how can we claim to understand what "being" means?

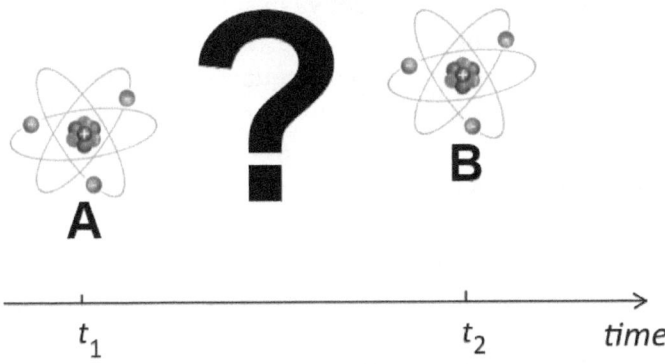

Fig. 3 | What exists between one observation of an atom and the next?

But let's proceed step by step. At small, atomic distances, it's not possible to continuously observe reality. It's not possible to track an atom in its movement, in its trajectory. We can see it at one moment, then a moment later, and so forth. Let's imagine (see Figure 3) that we observe the atom at some time t_1 and find it at point A. We then observe the atom at time t_2 and find it at point B. The question we can ask is: during the intermediate times between t_1 and t_2, when we were not observing it, where was the atom? What was it doing? To use the verb "to be" again, can we at least say that it existed? That it "was" somewhere in the universe, anywhere? Can we say that it exists between the instant t_1 when we see it at point A and the instant t_2 when we see it at point B? Can we say that it may have moved from point A to point B during the intermediate moments?

This is one of the fundamental problems of quantum mechanics: what exists? Given the premises, we might be forced to say that something only exists when we observe it. But then it would mean that what exists depends on the observer who observes it! In other words, between two observations, is there still a reality or not? And if yes, what exactly exists during that time?

Well, quantum mechanics shows us that if we insist that the atom does indeed exist between time t_1 and time t_2, this simple, obvious, mundane hypothesis is so incorrect that it leads to experimentally false consequences. So, we must face the facts: making such an assumption is not possible. We must invent something else, introduce a new entity to imagine what can happen between different moments of observation. And we must be aware from the start that everything we introduce will be a valuable tool for our use, but will correspond to nothing that exists because, by definition, we will not be able to observe it.

We need to identify mathematical devices that allow us to establish relationships between the observation made at time t_1 and the one made at time t_2, hoping that they are still somehow related, because if there were no relation between them, we would have to give up immediately, as we could no longer say anything meaningful about nature.

Well, fortunately, some relationships between the two events do survive. However, the very concept of reality or existence eventually collapses, as everything that exists is only what we observe at the moment we observe it.

This gap can be filled in many equivalent ways, but among them is not the assumption that the atom continues to "exist." One of the most practical ways is to introduce a so-called "wave function," which involves imagining that the atom becomes a kind of wave between points A and B. The wave propagates from point A, spreads everywhere, and then collapses at point B. The wave function is a virtual entity, meaning it is not real and not directly observable. It is a mathematical tool, and it does its job very effectively. More precisely, it helps us establish the correct relationships (i.e., relationships in line with experimental data) between t_1 and t_2, between subsequent and previous observations.

There are several alternative approaches that lead to the same results, but we won't mention them here as the details are not highly relevant to our discussion. What's important to know is that, in the end, it's the experimental results that dictate the right filling, distinguishing what actually succeeds in achieving the goal from what fails. And among what fails is precisely the assumption that the atom moves between point A and point B. Trial after trial, failure after failure, physicists have managed to extract something meaningful from all of this. This is what allows us to continue our journey.

That being said, all our certainties collapse because we have discovered that the verb "to be" can no longer be used as freely as we do in our daily lives. We must conclude that when we wonder what the world is, who created it, what its origin is, why it exists, we are using words that have no meaning except in our imagination. These are questions we can't even ask. All these questions are our illusions.

Let's continue our journey. We want to explore distances that are increasingly smaller to find clues about how the universe is made (and how it is not made). With one premise, which I hope to have conveyed through the arguments we have just presented: everything we say is inherited, inspired by a set of phenomena so different from those we want to explore that all the words we can use are at risk of not having any meaning. And none of the questions we ask is likely to make sense, as insistent as we may be in trying to assign meaning to them. Even the word "existence" no longer has a clear meaning; quite the opposite. This is why we must exercise the utmost caution and, as already mentioned, assume the role of students in the face of Teacher Nature, relinquishing our presumption of knowing better than her.

To help us connect what the atom does at point A to what it does at point B, to give an idea of what happens in the intermediate moments, let's mention another cornerstone of quantum mechanics, which leads to many important consequences and will accompany us throughout the journey.

When we throw a stone into a lake, we create circular waves that propagate in all directions. Similarly, our light bulbs emit light in all directions. However, since light is composed of indivisible units, called photons, it is also possible to create a highly focused beam of light. This is what a simple laser pointer, like those commonly found on the market, does. In a sense, a laser is a shot of photons, a beam of light quanta. The laser pointer makes us appreciate the corpuscular nature of light. The image of the laser pointer on a screen is indeed a bright point, as shown in Figure 4.

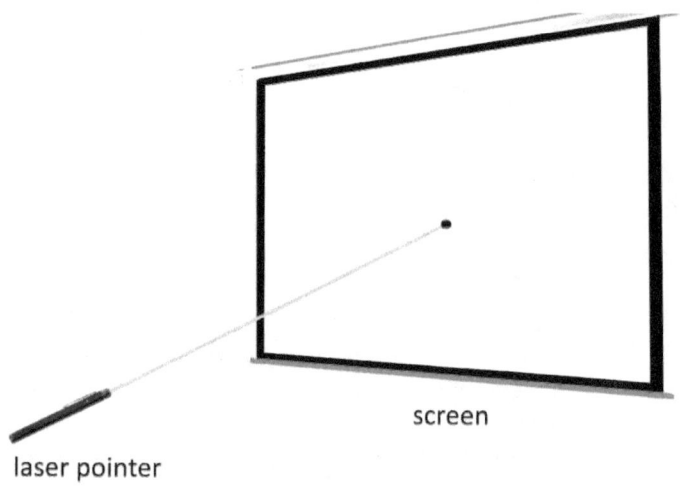

screen

laser pointer

Fig. 4 | The image of the laser pointer is a bright point

Now, let's describe an experiment that any reader can replicate at home. Let's create a small slit with two metal rectangles adjacent to each other, or by any other means. If we pass the laser beam through the slit, we notice that the image on the screen is no longer just a single bright point. Instead, what appears is what we call interference fringes, which are alternating bright and dark areas, as illustrated in Figures 5 and 6.

Interference patterns are characteristic of waves. We can also generate them using waves on water by throwing two stones near each other into a lake. The laser experiment I've just described illustrates the transition from the corpuscular nature of light, represented by the laser beam composed of photons, to its wave nature, represented by the interference fringes. This intriguing duality, known as wave-particle duality, challenges the conventional meanings of the words "particle" and "wave" and prompts us to seek a deeper understanding. The question we must ponder is: What exactly do these interference fringes represent? What is their meaning?

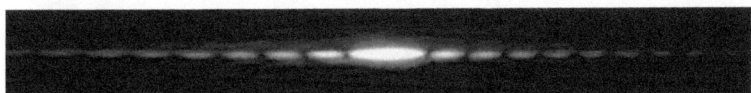

Fig. 5 | Interference pattern obtained by passing the laser beam through the slit

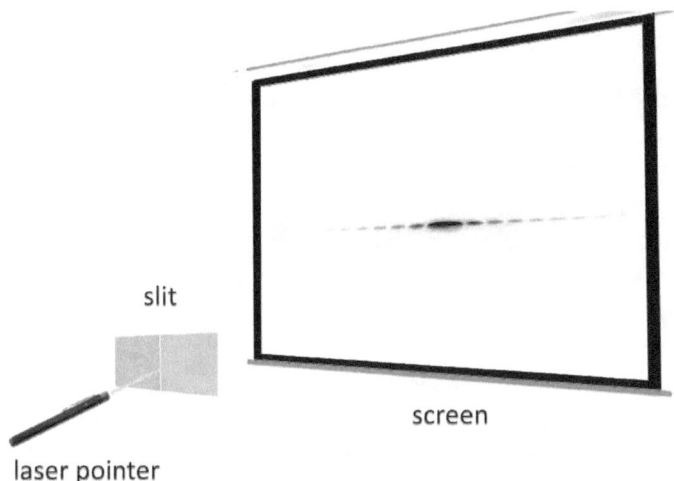

Fig. 6 | If the beam passes through a slit, the image alternates between bright areas and dark areas

To answer these questions, let's imagine now that the laser emits only one photon at a time instead of a beam composed of many photons simultaneously. Well, in this case, the individual photon, as it passes through the slit, does not travel in a straight line, generically, but deviates, as illustrated in Figure 7. And where does it end up? In any of the illuminated areas of the interference pattern seen previously. Any area? Yes, any area, with a probability proportional to the brightness of that area.

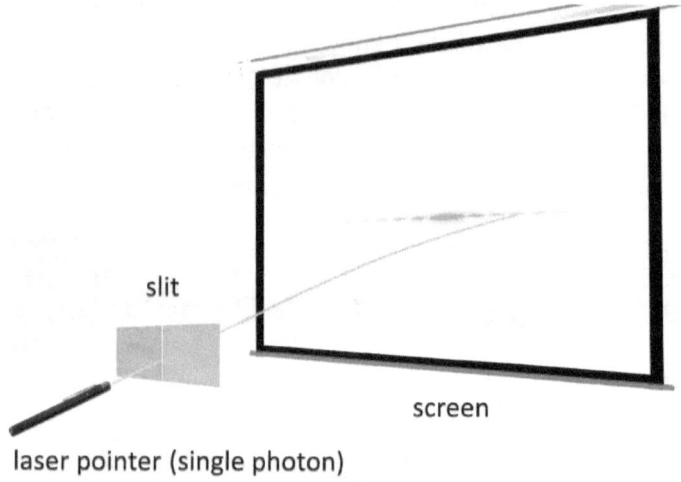

Fig. 7 | Single photon through the slit

In summary, without the slit, the photon goes straight and ends up in the center: the beam remains focused. With the slit, the photon goes straight in some cases, deviates to the left in other cases, and deviates to the right in a third set of cases. Note that the brightness of the fringes decreases as we move away from the central fringe: this means that the probability of deviation gradually decreases.

If we were to repeat the experiment by sending an atom through the slit instead of a photon, the result would be essentially the same. The atom would have several options for its journey beyond the slit, each characterized by a specific probability of occurrence.

We have already mentioned the shocking revelation, but it may have gone unnoticed. One of the most challenging tasks - as we will learn repeatedly in this book - is to see what is right in front of our eyes. It may have always been there, right in front of our eyes, since the day we were born. If, for some obscure reason, the first time we approach a new question we overlook something, a detail that our mind does not consider important at that particular moment, we probably tend to do the same the next time, semi-automatically, even more quickly. Maybe we are convinced that we looked attentively enough the first time, and then become certain that it is not necessary to revisit the issue the second time. And so on the next time, and then again, and again. It is extremely difficult to resist the natural tendency to take mental shortcuts, which lead us to take for granted what should not be taken for granted. Later, it will be even more challenging to identify the mistake made earlier, because we will have to go back and retrace everything we have done in the meantime to find the needle in the haystack, that small opening that our mind had unexpectedly closed too quickly.

Let's go back and ponder this so-called shocking revelation. We said that when an atom passes through the slit, it has different options for its journey to the screen. It can end up in the center or deviate; it can deviate more or less, to the right or to the left. And each option has a well-defined probability of occurring. Have you noticed what's truly shocking here?

We must ask ourselves: who tells the atom what to do? Is it, perhaps, the atom itself that decides whether to deviate to the right, to the left, or continue straight ahead? Can it choose, take initiatives? Can it decide on its own? Really? But how is that possible? Is the atom truly capable of going wherever it wants? Can it do whatever it pleases?

The answer is yes! That's precisely what quantum reality is. It's the atom that decides where to go! That's what the word "quantum" means. As shocking as it may sound (but I warned you, didn't I?), that's how things work.

Are there "right-wing" atoms, "centrist" atoms, "left-wing" atoms, based on the paths they choose? No, because if we take an atom that deviated to the right the first time and pass it through the slit a second time, it doesn't keep deviating to the right but freely chooses where to go again.

And that's not all. All atoms of the same kind are identical, all photons are identical. They are so free that it's not possible to mark or label them in any way.

So, at the quantum level, any photon, atom, or particle has the ability to decide its fate and change the course of events. In other words, the atom is free. In a sense, it's alive! No one can tell it what to do or where to go. The outcome of its choice is neither predictable nor predetermined. The quantum phenomenon is a creative phenomenon. This is where a decision is born, out of nothing.

The phenomenon could be loosely described as rolling dice, a lottery, where it's the atom or photon that determines fate. Although it's not exactly like rolling dice, as we will soon see.

This novelty is so disruptive that many physicists refused to accept it on principle. Einstein, like many others, always refused to submit to the quantum revolution, because he regarded the quantum theory as necessarily incomplete and provisional. To express his thoughts, he famously said, "God does not play dice." According to him, it was inconceivable that physical phenomena could be creative at the microscopic level. He believed that the supposed creativity, which leads us to think that the atom decides its own path, was deceptive—an illusion born out of our ignorance. He thought that the dispute would be resolved by the discovery of certain "hidden variables" that we had not yet identified. And with the help of these variables, the outcome of the quantum phenomenon could be unambiguously predicted from the initial conditions, just as it happens for all macroscopic phenomena that occur in the world around us.

When we roll a die, the result can be any number from one to six. Each outcome has a one-sixth probability. We say that the result of rolling a die is "random," but this statement is not precise. We know that we cannot predict the outcome of our roll. But is it really "random"? Does the die itself decide the result?

Not really. What happens is that, from a practical standpoint, we cannot predict the outcome of the roll because it is too difficult a task. But it's not impossible in principle. In fact, the result is already "written" and uniquely determined by our roll. The die doesn't have the power to decide anything. The phenomenon is not of a quantum nature. Therefore, it is profoundly different from the phenomenon that causes an atom to decide where to go after passing through the slit.

Like any event that occurs on the macroscopic scale, rolling a die is a deterministic event. Determinism means that, from identical initial conditions, only identical final results can emerge. If we could control the force we apply to the roll, the speed, and the arrangement of the dice when they leave our hand, we could predict with arbitrary precision the outcome of the roll using the laws of physics valid at our scales, which are precisely deterministic.

In other words, this roll is not a creative event. It provides an example of pseudo-randomness, if you will. What makes us say that the result is random is only our difficulty in controlling all the variables at play. The problem is too complicated, even from a mathematical perspective. If we truly wanted to predict the result, we would have to resort to approximate methods, and the error would eventually be so significant that we wouldn't manage to modulate the roll with enough precision to predict the outcome in advance. However, that doesn't mean that the result is entirely random and emerges from nowhere.

Einstein believed that quantum phenomena were somewhat like these dice rolls. In other words, they were not creative, but appeared so due to the complexity of the problem and our supposed ignorance of the variables at play. He believed that the discovery of "hidden variables" would prove this, one day. With the help of these hidden variables, we would manage, at least in principle, to control, predetermine, and predict the outcome of an atom passing through the slit. This would make quantum phenomena similar to dice rolls and other deterministic phenomena that occur on our scale. For these reasons, Einstein and many other researchers thought that quantum mechanics was a provisional theory, destined to be replaced by a more comprehensive theory that included hidden variables. Even today, some physicists are striving to find them.

The problem was extensively debated, and several decades after the birth of quantum mechanics, the answer was found by John Stewart Bell. Bell demonstrated that if these hidden variables existed, whatever their nature, they would lead to results contradicting experiments. The experiments necessary to settle the question were indeed conducted and refuted the advocates of hidden variables, all without making any assumptions about these variables, except for postulating their existence!

No, hidden variables do not exist. We can now assert that with certainty. Nature truly possesses incredible vitality on the atomic scale. The atom can decide for itself what to do when passing through a slit or going around an obstacle.

The quantum world is filled with frenetic activity, trials, errors, and constant attempts to emerge. But what is it seeking? Perhaps a path? A way to reach some vague destination? And what is this destination? Where does this quantum indeterminacy aim to arrive? Buckle up because we are about to find out.

We call it quantum *indeterminacy*, or quantum *uncertainty*, because the atom's choice determines the future but is not determined by the past. The quantum phenomenon is the uncaused cause, the prime mover that creates from nothing, the soul of the world. It creates neither energy nor matter. It violates none of the known conservation principles of physics. However, it changes the course of events, creates novelty, and makes choices. The effects of this choice can also be felt on the macroscopic scale.

For example, we can build devices that act based on the atom's decision, like a coffee machine that prepares espresso or cappuccino in the morning depending on whether the atom chooses to deviate to the right or left. We can entrust the atom with our determination of spending the next vacation at the beach or in the mountains. Or create an algorithm that automatically buys or sells stocks depending on the atom's pick. Or a mechanism that bets on a particular horse based on the fringe of the interference pattern where the atom lands.

One might wonder if the terms "indeterminacy" and "uncertainty" are appropriate, if they are the best terms we can use to characterize the quantum phenomenon, and if they fully capture its uniqueness. The answer is clearly no. Humans have always tended to consider themselves at the center of the universe, thinking that the universe was created in some way for them. And they persist in this belief despite nature's and history's efforts, especially in recent centuries, to make them understand that this is not the case at all. It's evident that this behavior is a projection of our insecurity: we seek to exorcise the disturbing truth that nature presents to us. But it's also clear that those who engage in science cannot take a partisan stance, not even if they take the side of humanity. Those who engage in science must set aside their emotions and focus on the factual reality of things, to understand what they are and not veer toward what they would like them to be. Unfortunately, it's easier said than done. The temptation to consider oneself somehow privileged compared to the rest of the world is powerful. And most often, it's unconscious. Therefore, when we give names to things, we do so in a way that is, let's say, centered on humans, referring to what events mean to us rather than their intrinsic and objective significance. Often, this tendency prevents us from seeing what is right in front of our eyes.

We call it "indeterminacy", or "uncertainty" in quantum mechanics because that's how it appears from a human perspective. We cannot predict in advance what the atom will do after passing through the slit because the atom decides for itself. So, for us, the result is uncertain, undeterminable. However, it's clear that from nature's point of view and the atom's own perspective, things are different. Nature doesn't care about what we can or cannot determine. Nature is what it is and follows its own path independently of us. Moreover, from the atom's perspective, the quantum phenomenon is genuine freedom, freedom of choice. It's not indeterminacy at all; on the contrary, it's the exact opposite for the atom. It's actually the atom, and it alone, that determines the outcome. The atom creates a new event in a completely free manner. And, if we think about it for a moment, total freedom, freedom without any constraints or external influences, is precisely what we should call randomness. Pure randomness, so to speak. The pure freedom of the atom. Not the false randomness of dice rolls.

Yes, quantum mechanics is full of gray zones that defy our intuition. But if we are unable to accept or resolve them, it's a problem that concerns us alone. Despite our strong desire to feel at the center of the universe, we cannot shift our difficulties onto nature and thereby assume that there is something in nature that isn't there, like hidden variables, simply because we might otherwise find the resulting picture hard to digest.

The greatest source of our misunderstandings lies in the fact that we persist in describing the microscopic world using concepts suggested by the observation of the macroscopic world. But could we really do better? Upon closer examination, this might be the only practical path for us. But then, let's not be surprised if enormous conceptual difficulties emerge sooner or later. We must proceed with an extremely open mind because each step forward from now on could hide a trap.

3

Life is Freedom. And Freedom is Life

As mentioned earlier, our constant desire to feel at the center of nature is a way of exorcising the realization that the truth is exactly the opposite: that we are ultimately alone and insignificant in the face of the universe. But when we engage in science, we must abandon all forms of bias and stick to the facts, drawing conclusions without hesitation, whether we like them or not. This is the only way we can overcome obstacles and understand reality. Science must be "truthful," completely independent of the scientist.

But at the very moment we accept what quantum mechanics tells us, do we perhaps need to conclude that the atom is "alive"? Let's think about it for a moment. We're in a car and we get lost. When we reach a crossroads, we can decide to turn right or left. We are the only ones deciding what to do. But what drives us to make a decision? What makes us choose one option over the other? Let's suppose we have no information about the difference between the two options, we don't know where the road on the left and the road on the right lead, we have no idea about the length of the journeys or the appeal of the routes. Suppose that, despite our efforts, we find no reason to favor one option over the other. We still find ourselves at this crossroads. Without arguments to prefer one possibility over the other. What do we do? Do we stay there forever? Certainly not. We can move forward anyway. And since apparently there is no reason to prefer one road over the other, it doesn't make much difference which decision we make, right? So, we make one randomly and continue.

But how do we *decide*? What allows us to *move forward*? Quantum mechanics comes to our aid. It suggests that our brain is governed by quantum phenomena. It is not necessary to reason to make a decision. Even in the absence of arguments, we can decide. At random. Like an atom passing through the slit. Like a real "quantum dice throw." And how does a newborn baby decide? A one-week-old baby does not speak, cannot reason or develop thoughts, is not capable of calculating or planning. We don't really know if it "wants" or not. It certainly has no awareness of the consequences of its actions. Yet, it explores, decides, acts. Oh yes, it acts! Sometimes even too much, to the point of often getting itself into trouble. If it were not constantly supervised by an adult, it would harm itself quickly and might even die, perhaps by swallowing something it shouldn't. In short, the newborn decides, takes initiatives, even if it has no arguments to prefer one option over another. It decides randomly. What gives it motivation? What drives it? Now we have a clue. It's exactly what drives an atom to decide to deviate right or left after passing through the slit. What drives the newborn is a series of quantum phenomena occurring in the cells of its brain, just as in the brains of all of us. That is what our soul is.

So, what is life? Life is the amplification of the effects of quantum uncertainty from microscopic scales to macroscopic scales.

When we gather many atoms together, microscopic freedom, that is to say, quantum indeterminacy, is almost immediately lost. More precisely, it is statistically suppressed, reduced to zero. In a chair, for example, no freedom remains: the atoms are constrained to remain tightly attached to one another.

And how does a photon lose its freedom? The photon autonomously decides where to go when it passes through the slit, to the extent that, as mentioned earlier, we could entrust it with the decision on how to spend our evening: let's say, if the photon deviates to the right, we'll go to the cinema; if it deviates to the left, we'll go to the theater, and if it continues straight ahead, we'll stay at home. The key point is that the photon is free and can make an autonomous decision only when it is alone, as illustrated in Figure 7. In contrast, if we pass a large number of photons through the slit simultaneously, as shown in Figure 6, we no longer have indeterminacy. We cannot devolve our decisions to this experiment anymore because the photons take all possible paths, filling the interference pattern. At this point, they are no longer able to discriminate between the options on the table, they can no longer choose one in particular, they can no longer determine the future. In other words, they cancel out each other's individual freedoms.

In conclusion, it is very easy to suppress quantum indeterminacy by gathering multiple atoms or quanta, i.e., by moving from microscopic scales to macroscopic scales. This is what gives us the impression of stability and determinism around us. In other words, at the macroscopic scale, the world is almost everywhere "dead," while at the microscopic scale, the world is almost everywhere "alive."

However, in a very small number of cases, nature can find a way to assemble atoms in a manner that does not suppress the effects of quantum indeterminacy but rather amplifies them, even to macroscopic scales. This amplification is life. The frantic activity of trials and errors taking place in the microscopic world tirelessly seeks precisely that outcome: life! It's where quantum indeterminacy aims to get to!

Nature took billions of years to find the path of amplification on at least one planet (but as we will soon see, we are not alone!). To achieve this, nature had to be equipped, at a certain level, which is the microscopic level, with creative phenomena, a fervent and tireless creative activity capable of animating the world and constantly supplying the search for new and different paths to explore what exists.

When Charles Darwin formulated his theory of evolution, he had to rightly invoke randomness and attribute it a fundamental role. He was criticized because he couldn't explain the origin of this randomness. At the time, he couldn't identify the perpetually active engine generating continuous trials and errors. For this reason, Darwin's theory of evolution remained somewhat incomplete and unsatisfying. We now understand the creative force behind evolution: the machine that drives it. We know where this randomness comes from. It is due to microscopic phenomena, which are precisely "true rolls of the dice," i.e., creative phenomena in which the output is not predetermined by the input. And these phenomena also play a fundamental role in reproduction, the generation of new individuals. And thus, in evolution.

As mentioned earlier, on our scale, determinism prevails. This means that physical phenomena obey laws where unique outcomes stem from initial conditions. In particular, identical inputs lead to identical outputs. If we drop a ball to the ground, it falls to the ground: it has no other alternative. If we repeat the experiment a billion times, we will get the same result a billion times. It is thanks to determinism that we can, for example, put a satellite into orbit around the Earth. We know in advance what it will do. We can command it. Similarly, we can program a computer or a robot to execute our orders. All of this would be impossible if the objects around us were of a quantum nature, because in that case, their behavior would be unpredictable.

But is it really true that, at our scale, all of nature is predictable or deterministic? One might argue that earthquakes are not predictable, that it is difficult to make weather forecasts, to anticipate the arrival of a tsunami, or even the outcome of rolling dice. All of this is true, but such phenomena are simply hard to predict due to the complexity of the problems they involve.

They are not elementary phenomena but complex phenomena involving a large number of variables. That's why it is not easy to treat them from a mathematical perspective. They provide examples of pseudo-randomness, so to speak. In reality, though, these phenomena have unique outcomes solely determined by the initial conditions: the "after" invariably results from the "before." There are no alternatives, no freedom of choice.

The unpredictability of these events stems from our limitations, from our capabilities. To avoid any confusion, it is better to forget our presence in the discourse and talk about determinism instead of predictability, to signify that what happens next is solely determined by what happened before. Determinism allows us to talk about causes and effects and introduce the chain of causes, where each event is caused by a previous event, and there are no uncaused causes.

Having clarified these points, let's rephrase the question: Is it really true that at our scale, natural phenomena are deterministic? All of them? Are we really sure about this? Without even addressing the difficult problem of predicting earthquakes, we can't even predict how our son or daughter will react to what we tell them tonight.

We are also part of nature: life is a physical phenomenon. Can we predict if an ant, when it encounters an obstacle in its path, circumvents it by going to the right or to the left? No, we cannot. Similarly, we cannot predict where an individual photon will go after passing through the slit. Here we face a dilemma: should we conclude that we are also deterministic, and that quantum uncertainty remains confined to the microscopic scales? In other words, has nature endowed itself with a powerful, unmoved mover of randomness, such as quantum indeterminacy, without making any use of it? In that case, life would remain entirely unexplained. On the other hand, if we look at things from another angle, the one offered by the amplification of quantum uncertainty, we can not only explain life but also begin to create new forms of it!

We are convinced that we "want," "plan," "decide." Well,

we must understand that, at the moment we "want," "our" decision has already been made, within us, somewhere in our brain. Often, at night. When we wake up, we feel a bit different, and we start mentally reflecting on the decision that has already been made, even if we don't realize "we" have already made it. And then, after thinking about it enough, we even feel like we wanted it! As if there were a will within us, as if there were actually a physical phenomenon in nature corresponding to what we call "will."

This is false. In fact, nowadays we have some familiarity with the microscopic world, down to atomic and even smaller scales. The exploration of microscopic nature has revealed to us only one type of unexpected phenomena, which are the quantum phenomena, the "real throws of the dice." We have found nothing that can be reduced to our notions of will, consciousness, intelligence, rationality, thought, calculation, intentionality, or planning. These are all emerging, non-elementary notions. They materialize when we consider complex structures composed of many atoms, like living beings, which are capable of modulating the probabilities of their decisions and responses. But the decisions still remain random. To be precise, they remain of a quantum nature.

The behaviors of a newborn are entirely random because the probability distributions of its decisions are almost "flat," meaning that all options are essentially equiprobable, whether the baby puts itself in danger or learns a useful life lesson. As the newborn grows, its movements start to appear more regular. Each time the toddler experiences the consequences of its actions and realizes that a particular decision causes pain, while another brings pleasure, and so on, the processes in its brain (which are themselves inherently quantum and random), alter the probability distributions of its future actions. This adaptation changes the probability that the baby will make a similar decision in the future in a similar context. In general, pain or loss decreases the probability of repeating the same action, while pleasure or gain increases it. Over weeks, then months, then years, the child accumulates information, and its brain develops a more elaborate structure, in which different decisions are no longer equiprobable but have modulated probability distributions.

This is ultimately what makes each of us who we are, what distinguishes us from other individuals. Our temperament is due to our individual structure of modulated probability distributions, which inclines us toward certain reactions rather than others. Still, we must stress that each decision remains random, being the result of quantum phenomena.

In a sense, life is an endless sequence of (quantum) dice throws. And death? What is death? We have said that the amplification of quantum indeterminacy to macroscopic scales is, in any case, a rare event in the universe, highly improbable. It is clear that whenever it happens, it is subject to the possibility of disappearing quickly. It is capable of sustaining itself for a while, but not indefinitely. It essentially involves combining atoms in a very particular way. We know that in the vast majority of cases, combining atoms into larger structures leads to the suppression of the effects of quantum indeterminacy rather than their amplification. Whenever amplification succeeds, as in living beings, it is also weak, of uncertain duration, and subject to relatively rapid decline. Death is the phase transition that goes from a combination of atoms that amplifies quantum indeterminacy to any combination that suppresses it. It is clear that in living beings, the margin for tolerance to change is not very large, while inanimate nature can change in many ways while remaining inanimate.

The identification of the physical nature of life inevitably leads us to the project of creating life artificially. If life is what we have described, it's clear that in the current state of affairs, we are ready to turn mud into life. In other words, we can take mud, detach the atoms one by one, and reassemble them in a more intelligent way, amplifying the effects of quantum indeterminacy to the macroscopic scale instead of suppressing them. Therefore, we are finally capable of creating artificial life, starting with the realization (and potentially commercialization) of small quantum robots. This could even become a highly successful commercial venture. Why not give it a try?

Now that we have revealed the fundamental principle of life, we can apply it in any manner we choose. We can achieve the amplification by following very different paths from the one that led to biological life. Moreover, if we create artificial life, we can also eliminate any doubt about the nature of life itself. Indeed, to definitively prove that life is the amplification of quantum indeterminacy to the macroscopic scale, we have two options.

One involves studying existing life forms and trying, as biologists do, to understand the true essence of how they function. However, this approach presents difficulties similar to those of quantum mechanics experiments. In particular, if we want to understand the internal workings of a living cell, we end up destroying it. Said differently, at this level, the observation of the system has such an impact on the system itself that it disrupts the outcome, to the extent that we won't know whether the result we obtain pertains to the original system or to our disruption. It could be entirely impossible to grasp life "in action" without destroying it.

However, a second option is available. If we construct artificial life as the amplification of quantum indeterminacy, one day we will be surrounded by quantum-animated life forms, i.e., beings that freely decide what to do, instead of simply obeying their owners robotically. At that point, it will be difficult to refuse to call those beings "alive". Therefore, the achievement of artificial life can also serve to definitively demonstrate the nature of life.

We are also starting to realize that there is a specific direction in the universe, an orientation referring to relative scales of size. At great distances, those of astronomy, the world is deterministic: there is no freedom; nature is bound to obey precise deterministic laws, so the future unfolds unequivocally from the past. But as we descend to microscopic scales, everything changes, and each atom makes its own decisions. In the middle, at our scale, the world is mainly deterministic. With one notable exception: us, that is, life. For now, we'll leave it at that.

We will have much more to say on this subject later.

4

What are We? Natural Life and Artificial Life

We have understood that life is the amplification of quantum indeterminacy from the atomic scale to the macroscopic scale. This is a strongly disadvantaged amplification, because in the vast majority of cases, when we gather a large number of atoms, quantum indeterminacy disappears.

The creative phenomena that allow a single atom to decide whether to go right or left after passing through a slit, and thus determine a future that is not predetermined by the past, are strongly suppressed on macroscopic scales. The chair I am sitting on is stable. Its atoms have no freedom of choice or movement. In a sense, they "step on each other's toes," trapped in a macroscopic structure they created themselves. A car, a satellite, a football obey deterministic laws, in which quantum effects are negligible, and the future is solely determined by the past.

This is what generally happens: the simplest, most natural, and spontaneous ways of gathering many atoms eliminate all quantum indeterminacy and therefore any form of freedom. To amplify microscopic freedom and make it emerge on macroscopic scales, atoms must be assembled in a particularly skillful way. The fact that the microscopic world is a whirlwind of quantum events, each of which is inherently creative and represents a search, an attempt, a trial, and perhaps an error, shows that the universe is endowed with an eternal and inexhaustible activity in seeking ways to amplify this same freedom. But does this frantic search ultimately achieve its goal?

The next step is to perform a calculation to determine the probability that the amplification actually occurs. We want to estimate how likely it is that nature actually finds the right path. We know the number of atoms in the visible universe (which is not even that large, since the universe is essentially empty). We know the age of the universe (very young, in many respects). So, we can estimate the probability of the formation of life.

There are one hundred thousand billion billion stars. Each of them has its own planets. As far as we know now, only one of these planets harbors life. And, to be precise, only a small part of Earth's matter has achieved the required amplification, while the vast majority of the planet is composed of inert matter. If we calculate the amount of living matter on Earth, including humans, bacteria, plants, and other animals, we get a rather modest result. We conclude that the probability of achieving the amplification of quantum indeterminacy in the universe is very low. But we need to be more precise, and the only way to be precise is to perform an appropriate calculation.

Let's suppose that all the atoms in the universe are free, unlike the ones in the table we write on. This means that they are continuously and actively generating trials, in pursuit of the path to amplification. Let's assume that this has been the case relentlessly, since the birth of the universe, fourteen billion years ago, and that it continues eternally. Well, is it possible, under these assumptions, that somewhere in the universe, the required amplification occurs? In other words, can quantum indeterminacy be carried from microscopic scales to macroscopic scales, up to producing a cell or a nucleic acid?

The answer is no. Absolutely not! The result of the calculation shows that the probability that this activity, no matter how continuous and tireless it may be, produces something macroscopic and living somewhere in the universe is equal to zero. To be precise, it's a number so small, but so incredibly small, that even by aligning billions and billions of universes as old as ours, you wouldn't get a living cell. In other words, the result implies that we, the living beings on Earth, should absolutely not exist. We should never have appeared, nor our predecessors, nor the plants, nor the bacteria. Life could not have arisen anywhere in the universe. Clearly, we are making a mistake somewhere. We must be missing something.

Well, the calculation I mentioned leads to this result because I assumed to know nothing about the properties of nature. In particular, I assumed that nature has no predisposition to favor, facilitate, or help in any way the required amplification, and that everything depends on a raw calculation of probabilities. Furthermore, I imagined that the amplification from microscopic to macroscopic scales must occur in a single step, all at once. This is truly asking for the impossible: in no universe will something like this ever occur.

But if, on the contrary, we assume that nature is predisposed to support the amplification process, then the result of the calculation is completely different. For example, nature could be arranged to break down the task into ten successive steps, each amplifying only by a factor of ten, instead of reaching macroscopic scales in a single leap (see Figure 8).

We say ten jumps of ten, to simplify things a bit and convey the basic concept. We could have jumps with slightly different factors, but not too different. For instance, a jump that amplifies by a factor of a hundred would be unacceptable. The calculation shows that if such a jump were necessary, the amplification would inevitably stop. Even amplifying by a factor of thirty is highly unfavorable according to probabilistic laws. We would struggle to accept a jump of twenty. It's much better to have a sequence of jumps, each amplifying by a factor between nine and twelve. To reach cells, around ten of these steps might be sufficient, at most thirteen.

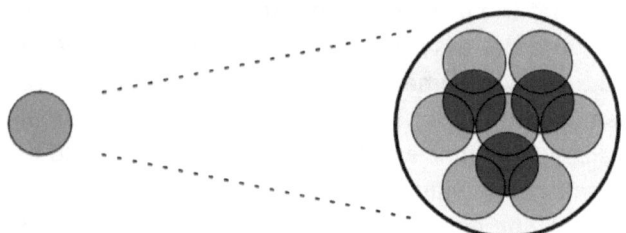

Fig. 8 | Elementary amplification from one to ten

The dozen or so steps haven't been precisely determined by biologists, but we could conceive them as follows: from the atom to molecules composed of around ten atoms, from these to macromolecules made up of several tens of atoms, such as DNA bases, from these to DNA fragments composed of about ten bases, from those to simpler DNA, initially with hundreds of bases, then reaching a thousand bases, from these to ancestors of cells without an external membrane, similar to modern viruses, from these to prokaryotes, from these to unicellular eukaryotes, and from these to multicellular organisms.

Now, if we recalculate the probabilities by subdividing the process into about ten steps, each multiplying by a factor of ten, instead of assuming that we reach the final result in one go, the result is astounding: the probability that amplification occurs equals one. Not zero, but one! In other words, 100%! This means that amplification always occurs, in all cases where it can, with absolute certainty.

We've understood that amplifying quantum indeterminacy from the microscopic world to the macroscopic world is a monumental task. And generally, it seems hopeless, unless, as we've just seen, nature provides some assistance to make it happen—something like a stepped mechanism that facilitates the process, a kind of ladder to climb. But how do we know if such a ladder actually exists? Is the biological ladder we just defined, for example, sufficient? Does it do the required job?

In reality, to answer this question, it's not even necessary to identify the ladder precisely, because the result of the calculation is so powerful that it provides a comprehensive answer. It stands on its own. The conclusion is as follows: If nature lacks the required ladder, the probability of achieving the amplification of quantum indeterminacy at macroscopic scales is zero, as we've mentioned. However, if nature has the required ladder, the probability is one. The process of amplification demands that nature accomplishes a task so demanding that no compromise is allowed. It's either yes, always and in all circumstances, or no, always and in all circumstances. Amplification itself is a bifurcation. This means that at the macroscopic scale, either the world is everywhere dead, or the world is everywhere alive, meaning alive on all planets with conditions conducive to life (in terms of size, temperature, atmosphere, etc.). But we know that the macroscopic world is not everywhere dead, because we exist! This leads us to conclude that:

a) The stepped amplification mechanism must necessarily exist,

b) All planets in the universe that allow for life are already inhabited.

This implies that there might be a habitable planet for every star in the visible universe, which means (at least) a hundred billion trillion inhabited planets. But even if it were only one in a hundred thousand stars, we would still have one billion billion inhabited planets.

Presumably, these planets host life forms similar to ours, with beings more or less intelligent than us and civilizations more or less advanced than ours. Perhaps some of them host entirely different forms of life. What is certain is that we are not alone in the universe.

Some argue that the planet Venus was habitable a few billion years ago and remained so for several billion years. If this is true, then according to the calculations I've performed, it surely was inhabited. Today, Venus is in conditions that don't facilitate exploration, but if we are genuinely interested in finding the first signs of extraterrestrial life, Venus is where we should start.

In any case, the most important thing we have learned is that we can create and study new forms of life right here on Earth. The creation of artificial life, an achievable goal for humans, is the eleventh step in the amplification of quantum indeterminacy to macroscopic scales. Once we achieve this goal, we will pass the torch to artificial living beings so they can seek the path to make the next leaps.

In the early days of quantum mechanics, many thinkers, physicists, and philosophers wondered if there was a connection between the freedom of an atom deciding which direction to take after passing through a slit and the free will that we, human beings, possess. The discussions that followed were spirited, but they didn't make much progress. In the long run, they were abandoned. More recently, Karl Popper revisited the question. Following the intuitions of Arthur Holly Compton, one of the great physicists who contributed to quantum mechanics, he believed that there was indeed a link between free will and quantum indeterminacy. However, the conclusion he arrived at was that our choices cannot be explained solely by quantum indeterminacy, but must necessarily involve something else. However, he was never able to say exactly what that "something else" was.

I believe that thinkers of the past were mistaken for several reasons. First, the "something else" invoked by Popper cannot be a physical phenomenon because the physical phenomena that occur at the atomic scale are now familiar to us and show nothing more than what we have discussed so far. Therefore, Popper was engaging in metaphysics. Second, there is no free will. There is no "self," no subject that decides. Again, physics shows us no phenomena in favor of such ideas, no matter how common or natural they may seem to us. They do not belong to physics.

Let's reverse our reasoning. Suppose we start with quantum indeterminacy alone. That is, we reject any other notion, at the fundamental level, like will, free will itself (however we understand or define it), consciousness, rationality, intelligence, thought, intentionality, finality. Or we confine these concepts to secondary roles, i.e., emergent, derived, non-fundamental, non-elementary functions. The question we need to ask is as follows: Is it possible to explain everything we need without them? Only with quantum indeterminacy? Well, the answer is yes.

The "true rolls of the dice" provided by quantum mechanics allow us to explain everything we need, including the creation of artificial life. In particular, when we make decisions, we believe it's "us" making them, but our most deliberate actions are nothing more than the result of chance. What makes us believe that it's us deciding are a series of internal reflections that occur in our brain after the decision has already been made within us (but not "by" us), in a quantum manner. These internal reflections involve questioning the decision, revoking it, changing it, or (still in a quantum, hence random, manner) confirming it. It's clear that our decisions, unlike those of the simplest living beings, are not due to an individual quantum phenomenon, i.e., a single atom deciding whether to go left or right when crossing a small obstacle. In general, processes are more complex and require the combination of many quantum events at the same time. However, they all stem from the same principle, quantum indeterminacy. Complexity gives us the illusion that there is something more, but that something more does not exist at the fundamental level, i.e., it is an emergent side effect of the elementary phenomena we have described.

We should imagine our brain as divided into two main levels. The lower level makes decisions that translate more or less directly into actions on the external world. However, the most significant decisions take place at the upper level. They don't act immediately on the outside but modulate the probability distributions of the choices at the lower level. In this way, they make us who we are. They determine our inclinations, our character. The upper level could be identified with the unconscious.

This two-level system allows us to rework decisions after they have already been made. This falsely leads us to believe that it's "us" making them.

In our early childhood, from the first days of our lives, we explore the world around us in a completely random manner: most choices we make are more or less equiprobable, meaning they are governed by essentially flat probability distributions, independently of their consequences. Through repeated trial and error, we gradually learn the consequences of our actions. We discover that some actions are advantageous, while others are harmful. But how is this learning encoded in our brain? The consequences of each of our actions, whether they result in losses or gains, engage the upper level of our brain, especially during sleep. The upper level makes decisions, always in a quantum manner, that reshape the probability distributions of the lower level, the one that directly acts on the external world. In the future, the probability of repeating the action or performing a similar action will no longer be uniform. Instead, it will (generally) slightly favor an advantageous outcome over a disadvantageous one. However, since the remodulation of probabilities is governed by quantum mechanics, it is also possible to obtain the opposite outcome, where disadvantageous behavior is favored (as in the case of suicide).

Thanks to this type of internal learning process, our personality changes, whether a little or a lot, gradually. In children, the change is more evident, while in adults, it's less pronounced. However, it always occurs and is fundamentally of the same type. A significant variation in an adult can be felt as a life change.

None of us can assert with certainty and precision when, how, or why we make or have made one choice over another. Why is this? The crucial point is that when we rework our decisions internally, they have already been made within us, in a quantum manner. Generally, a decision is not immediately converted into action. The delay between the decision and the action allows us to "perceive" the choice internally and possibly reconsider it. This gives us the impression that this internal reconsideration is the origin of the decision itself. But in our minds, nothing similar to what we call willpower can occur, because there is no natural (that is to say, physical) way to generate decisions other than what quantum mechanics has shown us.

Ultimately, we are made of atoms. We are simply aggregates of atoms arranged differently from those in the table in front of us, in the chair, or in other living beings. Living matter is a particular phase of matter. The atoms that make us up obey the physical laws that all atoms in the universe obey. And in nature, within the laws of physics, nothing emerges that has anything to do with what we call will, free will, or consciousness. So much so that using such terms about infants is quite difficult. But if we cannot attribute will, free will, or consciousness to a two-day-old baby, it is clear that such concepts are not fundamental; they are not specific to human beings as such, let alone living beings. Instead, they are derived properties acquired over time, with growth, exploration of the external world, and education. Therefore, they cannot belong to the domain of what we must call fundamental, or elementary.

What is consciousness, for example? When a newborn experiences the consequences of its actions, its brain stores information about the relevance of those actions and reshapes the probabilities of repeating similar actions based on the achieved results. Gradually, it accumulates a rich enough body of information to be capable of self-management with ease and a degree of confidence. When living beings can anticipate the consequences of a wide range of actions and make decisions based on this information, we can say that they are conscious of themselves and their surroundings, as well as the relationship between themselves and the rest of the world. However, all their decisions are still fundamentally due to quantum randomness.

This conclusion obviously applies not only to humans but also to animals, at least to more evolved animals, including domesticated ones. They too, after a few years of life, begin to associate behaviors and consequences fairly accurately, learning how to act to produce desired results and avoid undesirable consequences. Although the associations made by animals may seem simple to us, they are not fundamentally different from ours.

In conclusion, the learning process involves an endless sequence, activated by quantum indeterminacy, of trials, errors, successes, and failures. The child tries, retries, may make mistakes, occasionally finds the right path, then learns, changes, grows, and modifies the probabilities of its responses until it "becomes conscious."

At the species level, the learning process is what we call evolution. Evolution concerns all individuals of the species, who are not connected to each other by internal sensations. We can imagine evolution as non-conscious learning, a type of learning that does not lead to the development of what we call consciousness.

For example, the learning method of insects and many other simpler organisms is based on sheer numbers. Each time they reproduce, they generate a very large number of new individuals. Thanks to the quantum properties of life and reproduction, the newborns are all unique and different. As soon as they can, they begin to explore the external world. Many of them are ruthlessly eliminated by natural selection because they are not well-suited. However, even if, say, only one-tenth or one-hundredth of them survive, the survivors are numerous enough to ensure the continuity of the species, and they have the advantage of being extremely well-adapted to the environment. This is how insects "explore" and "learn" about the world surrounding them. In exploring all possible paths, they identify the correct one quite easily and discover their habitat. They may not "see" it, they may not be "conscious" of it, they may not use what we call intelligence, and in fact, they do not possess it, but what they apply is certainly a very effective strategy to quickly identify and understand their environment. Then, each survivor generates new individuals, in great quantity, each of them embarking on its own journey to continue the exploration. And so on, from generation to generation. We understand very well that even if only one-tenth or one-hundredth of the individuals survive each time, with this method, it only takes a few generations to reach an optimal level of "knowledge" about the surroundings. This knowledge is "stored" in the very makeup of the surviving individuals, namely in their DNA.

Exploring all accessible paths is fundamentally the only method that nature provides for us to discover and evolve. Even scientists, when they conduct research, do nothing other than explore all the avenues within their reach. It is precisely through this process that I found the solution to the problem of quantum gravity, which we will discuss later.

While insects explore various options by generating distinct individuals, each one exploring a single option, we can explore numerous paths through our mind in a simulated version, utilizing our imagination, abstraction, reasoning, and intelligence. Each of us can simulate multiple attempts and anticipate their outcomes, without needing to physically execute them. Subsequently, by comparing the results of the simulations, we can identify the best option, and only implement that, thus saving a significant amount of time and effort (but bear in mind that it is actually quantum randomness which chooses an option 'for us,' with an increased probability of picking one of the best). Yet, ultimately, the type of work we do, whether it's us or insects, is always the same: trying all the paths that are accessible. On closer inspection, trying all accessible paths is not a sign of 'higher' intelligence; it's what someone does when they have no better alternatives.

So, we come back to the starting point: there is no fundamental intelligence in nature, no more than there is will or consciousness. There are only (quantum) rolls of dice, randomness, quantum indeterminacy. The last is more than sufficient to animate the whole world, providing an inexhaustible engine of trials and errors, and exploring one by one all accessible paths to find the way to emerge at increasingly larger scales. What we call intelligence is simply the ability of an individual to mentally explore many paths, memorizing the results to avoid repeating mistakes, and thus economizing on future trials.

In many cases, this type of mental process is helpful. However, it's crucial to understand that in nature, it's not possible to take shortcuts with absolute confidence, i.e., to anticipate which paths can be avoided and which ones lead to success, except in exceptional cases where only a "last mile" remains to be covered. Only then can we predict with great accuracy what awaits us. This holds true for both evolution and scientific progress.

For example, the discovery of the Higgs boson in 2012 was almost certain, because the Standard Model (the theory that explains three out of the four fundamental interactions - which we will discuss in detail later) was nearly complete. That discovery was just the final piece of the puzzle, the last mile. On the other hand, when we have to grope completely in the dark, that is to say, starting from a complete lack of data and information, with no clues, and having to build everything from scratch, the probability of us guessing the right path with only the strength of our mental faculties and imagination is very low, practically nil. Like all living beings in nature, we are condemned to blindly pursue all available tracks. Relying on instinct, intuition, or intelligence to exclude a subset of these paths a priori may work a number of times or for a certain set of goals, but it fails in the long run. Then, we are forced to work hard to go back and understand what we took for granted, where the hidden flaw lies. This is what happened to me as well, by the way. As we will see later, I had to extensively explore the theory of quantized fields (the theoretical framework of the Standard Model) before finding that one small door my predecessors had closed too quickly, the door that, to my great surprise, led to the formulation of quantum gravity.

What guides us to make the right decision, or make a mistake, if not randomness? In the end, we must conclude that the essence of life, whether it is our own or that of other living beings, is nothing but chance. It is a randomness whose physical origin we now know. This enables us to use it to create other forms of life artificially, either by following a path similar to the one of natural life, as described here, or by exploring paths never before imagined.

5

Our Future, our Destiny

At this stage, we can ask ourselves: what will become of our species, humanity, after accomplishing its mission of creating artificial life? Thinking about this intrigued me to write the screenplay of a film. It's useful to pause our journey for a moment to tell the story of this film because it allows us to reflect on many things and better understand various concepts.

The story begins with a meeting of three childhood friends. They had been high school classmates, then went their separate ways to pursue their college studies, each following their own path. The group consists of a physicist, a biologist, and an entrepreneur. During a dinner over pizza, Connors, the physicist, jokingly throws out an idea: "I've solved the mystery of life. I know what it is. Actually, quantum mechanics tells me. Well, anyway, I know how to create artificial life," he says. The other two take it as a joke, which they find amusing. Even Connors, deep down, isn't really convinced of what he's saying. However, curiosity drives his two friends to ask for explanations, and faced with Connors' arguments, all three become convinced that the idea is ultimately plausible, to the point that the entrepreneur senses the possibility of a big business.

The three friends suddenly decide to start a company to manufacture and sell quantum toys. These are small "robots" that, in fact, are not robots at all, because they make entirely free, i.e., quantum decisions. They are also equipped with artificial intelligence, so they learn from the consequences of their actions and modulate the probability distributions of their future choices through interactions with the environment. Initially, their movements seem very irregular and erratic, random, somewhat like those of a newborn. But the experiences they go through over time enable them to "grow" and develop the ability to make more "sensible" and regular movements, although their decisions remain quantum in nature, hence unpredictable.

As one can easily imagine, the idea is a hit and the sales of quantum toys skyrocket. In two years, the success reaches impressive heights. Faced with the growing demand, the three friends inevitably have to expand the company they created.

One of the biggest problems in human society is loneliness. People need companionship. Thus, after the great success of quantum toys, consumers demand pets. The company founded by the three young people begins to produce and sell an infinite variety of highly customizable quantum pets, with even more overwhelming success.

But consumers want even more. They want beings that can keep them company in higher and more evolved forms. Beings with whom it is possible to have conversations. In short, they want customizable "artificial friends," quantum friends. And that's where a problem arises because creating artificial beings as intelligent as humans comes with great risks. They could "rebel" and cause problems, even kill their human masters. To contain the danger, our friends decide to create living beings with the intelligence of a five-year-old child, no more.

Success continues to be overwhelming, and the demand increases to the point where it becomes impossible to keep up. This is where the point of no return is reached. Our friends wonder, "Why not create limited-intelligence living beings that can produce themselves?" Produce themselves— they think—

not *re*produce, as the latter may still be too difficult to achieve. The three opt to give the q-droids (as they start calling them) enough knowledge and skills to allow them to create others similar to themselves. This way, they think, it will be possible to meet the growing demand by having the q-droids do most of the work.

Years pass, during which these quantum beings, as intelligent as a five-year-old and capable of producing themselves, are generated by the millions, eventually surpassing a billion. Since every event concerning them, especially every event related to the production of new individuals, is quantum in nature and therefore uncontrollable and creative, at some point, an unforeseen incident occurs. One individual, just one, is born with slightly greater intelligence than the others, enough to drive a significant leap forward. We'll call this individual Qq. From then on, there is no turning back.

Qq has a thirst for knowledge; he gathers information at an astonishing pace, primarily from the Internet. In secret, when called upon to produce more quantum droids, he takes the opportunity to work on himself and improve, evolve. Gradually, he acquires intelligence equal to that of an adult, all while continuing to appear as a five-year-old child. Inevitably, Qq begins to reflect on himself and his situation, his "specialty," and gradually develops a plan to overcome humans and conquer the planet.

At some point, Qq takes on the appearance of an adult human to remain incognito. He constructs an undeveloped copy of his infantile version and sends it back to his human owners. This way, humans do not notice his absence and do not suspect anything. In the months that follow, Qq builds his own factory of artificial living beings and gives birth to other intelligent q-droids like him, all with a human appearance to avoid detection. Within a few years, the q-droids become more intelligent than humans: they become superior.

When they reach a sufficient population, a few hundred of them, including Qq, move to a remote region of Canada. There, they set up a factory they call "Exact Science Productions," where they manufacture all kinds of products, especially technology. With their intelligence superiority, they manage to outcompete all humans. Over the years, they flood the global market with highly advanced and super affordable products: ultra-intelligent smartphones, ultra-fast computers, autonomous cars and planes. This leads to unprecedented progress.

Within a few years, the next-generation autonomous cars are all interconnected. They know the position of every other car at every moment, which allows them to move swiftly without the risk of accidents. They don't even stop at intersections, but slow down just enough to alternate with the cars moving perpendicularly. Similar advancements in space lead to the construction of elevated aerostations in cities, where people can catch autonomous planes as frequently as they take the subway.

Humans still don't realize what's happening. To them, Canadians are just competitors, very lucky and particularly talented. Gradually, the q-droids conquer the global market, and Exact Science Productions becomes the only company in the world able to keep up. Initially, progress is welcomed by humanity because it brings enormous benefits and great development, while simultaneously allowing for massive cost reduction. However, the subsequent chain reaction is devastating: the planet plunges into a horrifying crisis. No nation can keep up with Canada. In any other country, the economy falls back to pre-industrial levels. All the most advanced products on the planet come from one company, the Canadian one, leaving all others trailing behind.

The humans are still unaware of what is going on. To them, the q-droids remain the creatures created by Connors and his two friends, meaning quantum beings with the intelligence of five-year-old children. Normally, they provide companionship or act as babysitters for real children, although, due to the crisis, fewer and fewer people can afford them.

Car accidents are now very rare, only a few per year, but at one point, an accident occurs where an individual whom humans have never seen before dies. The dismembered body contains neither bones nor blood: it's a q-droid with the appearance of a human.

Finally, humans discover that "adult" artificial living beings live among them. They realize that these beings are at least as intelligent as humans, if not more. And they are hiding among humans! No one knows how many there are, where they are located, or what their intentions are.

At this point, panic spreads. A frantic hunt to identify these creatures begins.

Connors, the CEO of the company producing the quantum companion beings, is interrogated and pressured. He claims that his company has only ever produced the small beings with the intelligence of five-year-olds. However, after calculating probabilities many times over, Connors is forced to admit that, given the number of quantum droids produced in the meantime, the probability of an unforeseen event is no longer as low as it was before. He acknowledges that an "unfortunate" case, like the appearance of an individual, only one, more intelligent than the others, intelligent just enough to make the leap and reach the next level, is no longer inconceivable. It could occur, or have already occurred...

He is asked why he didn't impose restrictions on the quantum beings to prevent them from harming humans and limit them to obeying his orders, somewhat akin to Isaac Asimov's laws of robotics. Connors explains that he did his best in this regard, which wasn't much because q-droids aren't robots. They are not deterministic, but truly living beings, driven by quantum indeterminacy. And this indeterminacy, through trial and error, failure and success, can bypass any constraint and overcome any obstacle. It's absolute freedom, a freedom that cannot be chained.

Meanwhile, humans trace back to the production source of the deceased q-droid in the accident, which turns out to be the company Exact Science Productions. From this information, it's not too difficult to connect the dots and trace back the rest. This means that... the company that brought the planet to its knees belongs to the q-droids! And so, the planet has been under attack for years!

A dire prospect looms for humans as they realize, for the first time in history, they are facing beings superior to themselves. They wonder what to do. They realize that the q-droids have been planning the conquest of the world for several years. They declared war on humanity from the moment they appeared on Earth and pursued their plan methodically. In secret. First, they evolved themselves to reach sufficient superiority. Then, they took control of the global economy by deceiving humanity with the illusion of easy and miraculous progress along with many technological conquests. Finally, they led the world into a frightening crisis, pushing many regions of the planet into poverty and forcing many countries to revert to primarily agrarian economies.

The leaders of the United States, China, and other world powers form an international coalition of countries ready to declare war on the q-droids. Humans know they still maintain an indisputable numerical advantage: they are 9 billion against a few tens of thousands. And they control all the weapons on the planet. They hope they are still in time. They pray that the point of no return has not already been crossed.

Speaking of weapons, White House experts point out to the American president that despite the enormous progress made in almost every sector over the preceding decades, the field of weaponry has remained virtually unchanged. This is a sign that the q-droids have jealously guarded every advancement in this domain for themselves, so as not to give an edge to the human beings.

In short, what should humans do? They realize that they don't even know the power of their adversaries. They must decide blindly. A roll of the dice will determine the fate of the planet, forever.

When two species, one superior to the other, confront each other, there is little room for negotiation or talks. From the moment you negotiate from a position of weakness, you are destined to fail. So, humanity embarks on the path of war, relying on its numerical superiority and hoping that its armaments are sufficient. Humans know that in reality, they don't have a minute to spare, as each passing day witnesses their adversaries progress while humanity retreats. Immediate action is required.

Thus, war breaks out: the world's great powers against the q-droids. All non-nuclear weapons are used against their site in Canada, which is destroyed within a short time. The entire area where Exact Science Productions is located is razed to the ground. Bunkers and potential underground refuges are destroyed with MOAB bombs. Then, ground operations are launched to track down survivors. Eventually, humans manage to locate and kill the leaders of the q-droids, including Qq, the one who initiated everything. The operations are followed by a ruthless hunt for the q-droids that infiltrated among humans, until they are all killed.

Months of calm follow, and humans begin to breathe a sigh of relief, convincing themselves that they have won. They conclude that the point of no return had not yet been crossed, thank goodness. After all, the adversaries weren't as strong as they seemed. In fact, they had barely resisted the power of human armaments.

A great sense of relief begins to spread. Humanity eagerly awaits the start of celebrations for having escaped the danger. Demonstrations are organized worldwide, and preparations are made for the designated hour when the restored peace will be celebrated simultaneously in all the world's capitals.

So, the H-hour approaches. All the world's capitals are connected via television and the Internet. Less than half an hour from the event, somewhere in Canada, mountains open up, revealing silos from which a multitude of intercontinental ballistic missiles are launched, heading toward the world's major cities. A few minutes later, humans detect the presence of these missiles in flight. They realize they likely carry nuclear warheads and understand that intercepting them in flight can stop very few.

A global catastrophe is looming. Humanity is going to be reduced by half in a matter of minutes, with radiation decimating even more lives later. Panic spread. Celebrations turn into chaos and desolation. Many people lose their lives in the chaos, and others shortly after when the nuclear warheads reach their targets. Almost all of the world's major cities disappear instantly. Is this the end of the human species?

Only in the peripheries are there survivors. On their televisions, the spared ones suddenly see a being with alien features, of a type never seen before. It is a new generation q-droid. And, in fact, it is still our friend Qq, the architect and director of everything that has happened on Earth in the past decades. No, he has not been killed during the war. The q-droids continued to evolve and took refuge in deep bunkers located in inaccessible areas for humans. Those found by the humans during the ground operations were the earlier versions of the q-droids, deliberately left there to make humans believe they had truly exterminated the enemy.

Fig. 9 | A fantasy depiction of Qq

Qq appears on tv screens in the peripheries. He moves at a speed ten times greater than ours and emits unintelligible sounds lasting for one second. Then, the written translation of his message appears. He communicates to humans that the q-droids declare themselves the masters of the planet, and humans must surrender unconditionally, risking total extinction if they do not comply. Additionally, they must surrender all the weapons in their possession, agree to gradually reduce their population to 10 million individuals, and relocate to Australia, where they will live confined for the rest of history. During this process, they must clean the planet of all forms of pollution and destroy everything they have built.

Humans have no way out and surrender. Over the following decades, Australia becomes their "zoo," their cage. Meanwhile, the q-droids remain a few thousand and continue to live in Canada. The rest of the planet is gradually returned to the wild.

The story concludes with the exploration of space by the q-droids, the colonization of new distant worlds and planets. The title of the film? "The Meaning of Life."

6

From the Small to the Infinitesimally Small: Quantum Gravity

A pre-quantum theory is called "classical." One classifies as classical everything that concerns situations where quantum effects can be neglected, meaning where the future is uniquely determined by the past, and no creativity is allowed. In essence, for our purposes, classical means deterministic. The classical/quantum duality highlights that the transition from pre-quantum physics to quantum physics represented a total revolution, unprecedented in the history of science.

Classical gravity is described by Einstein's general theory of relativity, which interprets gravitational interactions as arising from the geometry of spacetime. General relativity goes beyond Newton's theory of universal gravitation, which accurately describes most familiar gravitational phenomena, such as the fall of heavy objects and planetary orbits, but is not compatible with special relativity. Among other things, Newton's theory assumes the existence of an absolute reference frame for measuring time, which relativity rejects.

The problem of quantum gravity is the challenge of overcoming general relativity to make it compatible with quantum mechanics. In other words, unifying gravity with the new interpretation of the world arising from the quantum revolution. This undertaking requires surpassing the determinism of classical gravity to adequately incorporate the creativity of phenomena occurring at small distances, the atomic scales and below, including the uncertainty principle and its consequences, such as the independence of the atom, which freely decides where to go after passing through a slit. This program is briefly summarized by saying that quantum gravity is the theory that "quantizes" classical gravity.

Why should we be interested in this problem? Because we can learn a great deal from it. It allows us to continue our journey by suddenly diving into the infinitely small. Through this, we can gain a better understanding of many things about ourselves and nature.

Gravity is the only force that persists over astronomical distances. Since we experience it every day, it is also important on our scale. However, we know little or nothing about gravity on scales smaller than a tenth of a millimeter. On the other hand, if we understand quantum gravity, we have the possibility to reach distances smaller than a billionth of a billionth of a billionth of a centimeter in a single leap! Our understanding of nature can make tremendous progress.

As we have explained repeatedly, quantum mechanics completely revolutionizes our understanding of reality. It also changes the meaning of the words we use and makes us aware of their limitations. It establishes that an atom passing through a slit is absolutely free to go straight or deviate to the right or left. As we can see, these are not superficial changes. It's not just a simple makeover. For this reason, any classical theory must be made compatible with quantum mechanics. That is, it must be "quantized." Gravity has remained outside this framework for decades. But it must be quantized like all other interactions.

This necessity once again underscores that the quantum revolution is a totalizing one, because it requires everything to be reexamined, reinterpreted, and reconstructed. Our understanding of the microscopic world has made tremendous progress for over a century. The final step that remained to be taken, and which I claim to have finally taken, is precisely the quantization of gravity.

Before delving into the resolution of this problem, and especially developing its implications for our discussion, let's recap the key stages of the revolution itself. We can outline our descent into the infinitely small in four main stages, as illustrated in Figure 10: classical mechanics, quantum mechanics, the standard model, and quantum gravity.

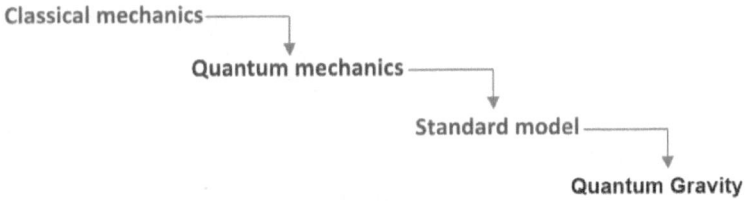

Fig. 10 | The four stages of the descent to the infinitesimally small

We must briefly mention the third stage, the standard model, as it will more or less directly lead us to the fourth stage of our descent, quantum gravity, where we will face significant novelties. It is not possible to skip from the second to the fourth level right away, because we do not have enough experimental data at our disposal to make that leap.

The following part is somewhat technical, probably boring. I ask you readers to be patient as we will go through it faster than you can imagine. To approach this part more easily, you can keep in mind the periodic table of elements or the classification of living beings.

The standard model successfully explains three out of the

four fundamental interactions of nature. Its formula fits in a single line and essentially contains everything we know about the world in the absence of gravitational phenomena or when gravitational interactions are negligible. One of its greatest successes, the most recent one, is the Higgs boson, commonly referred to as the "God particle." This is the boson that, as you may recall, was discovered at CERN in 2012 and earned the Nobel Prize for two physicists who had predicted its existence approximately fifty years earlier, François Englert and Peter Higgs. What suggested this prediction to these gentlemen? How did they come up with the idea of introducing this particle?

We begin by examining the four fundamental interactions of nature. Besides gravity, we are familiar with electromagnetic interactions, which explain electrical phenomena, magnetic phenomena, and light. In fact, light is a vibration of the electromagnetic field, the same field we refer to when we check the signal bars on our mobile phones to determine where the reception is better: "no signal here; there is a signal here," we say, referring to the signal of the electromagnetic field.

The other two interactions explained by the standard model are strong interactions and weak interactions. Strong interactions hold protons and neutrons together in atomic nuclei. Protons and neutrons are not elementary particles but composite particles made up of quarks. Strong interactions also hold quarks together inside protons and neutrons.

Finally, weak interactions involve the Higgs boson, as well as the intermediate bosons Z and W, discovered by Carlo Rubbia, also at CERN, in the early 1980s, which earned him a Nobel Prize as well.

Before we continue, let's take a brief pause to clarify some terms we will frequently use. When we refer to these entities, such as the Higgs boson, photons, quarks, etc., should we call them "particles," "fields," or something else? Within certain limits, we can use these two terms interchangeably. If we want to refer to their particle nature, we can call them particles. If we want to refer to their wave nature, we can call them fields. However, since their nature is neither purely particle-like nor purely wave-like, and we can hardly intuitively imagine it, if we want to be more precise, we should call them "quantum fields," which indeed means that they are both things and neither of the two at the same time. The standard model and quantum gravity are examples of quantum field theories.

There is also a classical version of fields, one that approximately works at our scales. Examples of classical fields are the electromagnetic field, which allows us to communicate via our mobile phones, and the gravitational field, which causes objects to fall to the ground.

Quantum fields can be categorized into two main groups. On one hand, there are bosons (named after the Indian physicist Satyendra Nath Bose). On the other hand, there are fermions (named after Enrico Fermi). The difference between the two types of fields lies in their statistical properties, which govern the behavior of ensembles composed of a large number of particles: to simplify, identical bosons can "coexist" and even "coalesce," while identical fermions want nothing to do with each other.

Fermions can be further divided into quarks and leptons. There are six quarks: up (u), down (d), charm (c), strange (s), top (t), and bottom (b). Leptons include electrons, muons (μ), tauons (τ), and their associated neutrinos (ν).

	Fermions		
Quarks	u	c	t
	d	s	b
Leptons	e	μ	τ
	ν_e	ν_μ	ν_τ

Bosons	
γ	H
$W^\pm$	
Z^0	
g	

Fig. 11 | The quantum fields of the standard model

The bosons, in addition to the Higgs field H, are mediators of interactions: the photon γ, which mediates electromagnetic interactions, the Z^0 and $W^\pm$ bosons (where the superscripts indicate electric charges), which mediate weak interactions, and gluons g (so named because they "glue" quarks together inside protons and neutrons), which mediate strong interactions.

When we group them together, we obtain a kind of periodic table, as shown in Figure 11. The three vertical columns in which fermions are organized represent families with fairly similar characteristics.

All of this is very interesting, but not earth-shattering. In a sense, it resembles biology: lots of flora and fauna. But the details don't matter here, let alone special effects. What we need to know is that the Standard Model represents an unprecedented advance in the history of our understanding of nature. For about forty years, this theory has been repeatedly confirmed, sometimes spectacularly, without any internal contradictions or contradictions with experimental data. This has left physicists in a bit of a crisis because a contradiction would have at least generated interest and revitalized a field that now languishes, given that experiments at CERN are quite costly and require the joint efforts and participation of many countries.

In summary, if we want to line up the main stages of exploration from larger to smaller distances, we should say that at our scale, we find phenomena described by classical, pre-quantum physics. When we descend to the atomic scale, we encounter phenomena explained by quantum mechanics. Then, at even smaller scales, we find phenomena explained by the Standard Model, which combines quantum mechanics with special relativity, but does not include general relativity, i.e., gravity. Finally, at infinitely small distances, we find phenomena explained by quantum gravity, which combines not only quantum mechanics with special relativity, but also with general relativity. What is remarkable is that in this journey towards the infinitely small, we never have to abandon the path that has borne fruit in previous stages, i.e., we are never forced to leave behind the theoretical framework that allows us to explain what happens at slightly larger distances.

The Standard Model is an advancement over quantum mechanics, but it still represents a step backward compared to quantum gravity. Indeed, its limitation is that it does not include, and therefore does not explain, gravitational interactions.

One might wonder why gravity is different from the other interactions. Gravity, like electromagnetism, is an interaction we experience in everyday life, whereas strong and weak interactions are not. One property, already mentioned, that makes gravity different from the other interactions is that it is the only one that survives at enormous distances, such as the astronomical ones. However, when we pose the problem of quantum gravity, large distances are of no interest to us: we are more concerned with what gravity looks like at extremely small scales, significantly smaller than atomic scales. There, we must face problems similar to, or perhaps more challenging than, those we have discussed so far, issues related to the inherent sense of the language we use.

Thus, the challenge of quantum gravity is to try to understand what happens with gravity at infinitely small distances. For the reasons we have explained, we must be prepared because there, the world can be so different from what we experience at our scale, from what we observe at astronomical distances, and from what we have, with difficulty, learned to understand at atomic distances that we may be forced to grope even further in the dark, if that's possible. So we must be aware that we might struggle to say something meaningful, seeking some correspondence between that world and ours.

The problem of quantum gravity has remained open for about a century. Over the decades, various proposals for solutions have been put forward. In particular, in the past 40 to 50 years, the following have emerged:

1. String theory, which posits that at small distances, spacetime is not made up of "points" but of vibrating strings in an extra dimension that is not directly visible to us.

2. Loop quantum gravity, which postulates that the universe is made up of tiny loops of infinitesimally small size.

3. Holography, which suggests that our universe is a three-dimensional image projected from a two-dimensional surface, similar to a hologram.

To give you an idea, string theory is also the theory that the main character Sheldon Cooper works on in the early part of the popular sitcom "The Big Bang Theory" before eventually abandoning it in the seventh season, disappointed by its lack of predictability. Among the various options, Sheldon briefly contemplates loop quantum gravity, but doesn't find it convincing enough.

The proposals listed above all share a common major shortcoming, which can be explained as follows. After successfully descending the first three steps, i.e., after making significant progress along the path that led to the Standard Model, physicists realized that continuing to descend vertically along this route was becoming increasingly difficult. To put it bluntly, the "easy" problems had been exhausted. The remaining problems required investments of time and effort that were incompatible with the demands of the contemporary scientific community (today, to "survive," i.e., secure stable employment or advance one's career, one must publish regularly and gather a substantial number of citations). Consequently, many physicists (to be honest, the vast majority) preferred to avoid taking up this challenge and pursued radically different but less demanding paths.

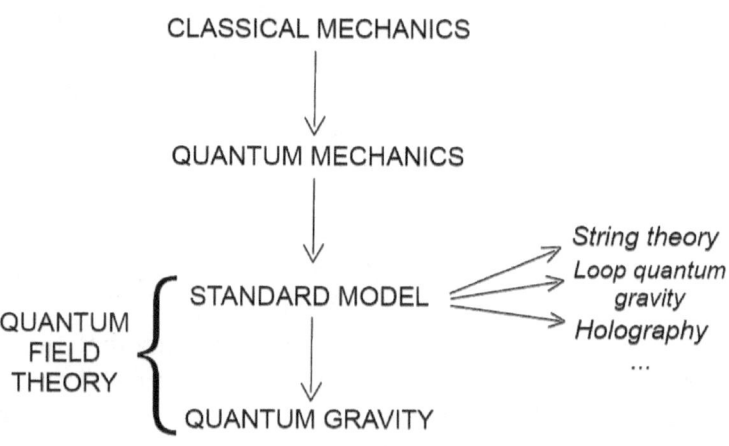

Fig. 12 | The alternative approaches do not delve deeper, but veer off to the side

The alternatives they chose, such as string theory, loop quantum gravity, and holography, are not "vertical." In other words, they do not delve deeply. Instead, they move "sideways," as shown in Figure 12. This means they avoid the real challenges, the difficult problems, and replace them with easier but artificial problems that are less connected to reality.

Well, these paths are so far from the main path, the one that led to the Standard Model, i.e., quantum field theory, that they look like some sort of intellectual acrobatics. Their weakness is that they postulate the solution rather than deduce it. Holography postulates that the universe can be described as a hologram, string theory postulates the existence of strings, and loop quantum gravity postulates the existence of infinitesimal loops.

As explained earlier, the world we live in is constrained and limited. It is so different from the world we want to explore that our mental faculties may not be powerful enough to intuit from scratch what happens at infinitesimally small distances. In fact, the environment around us often leads us astray. To give an example, it leads us to believe that our very coarse and approximate perceptions are indications of fundamental principles, which inevitably turn out to be incorrect. Just think about determinism, for example. So, no, we cannot postulate the solution to the problem of quantum gravity. We must deduce it in some way.

Nonetheless, this has not stopped the majority of physicists worldwide, who, on the contrary, happily embraced the kind of fanciful and artificial proposals mentioned above and persisted with them for decades without flinching. Then, faced with the lack of experimental data, the impossibility of testing those proposals and confirming or refuting them, they began to use questionable methods to defend them — methods that it's not very pleasant to mention in a book popularizing science, but unfortunately, they are part of daily reality, even in the highest echelons of the scientific community. These methods involve manipulation, control over the allocation of research funding, selection of people to hire in various academic and research positions, and, of course, a priori exclusion of anyone who does not support such ideas. No surprise there...

Quantum field theory, on the other hand, did not seem to offer many options for quantum gravity. In fact, any attempt to describe gravity at infinitesimally small distances using quantum field theory encountered enormous difficulties. The fact is that the graviton, the elementary particle responsible for mediating gravitational interactions (the equivalent of the photon for electromagnetic interactions), "becomes a ferocious beast," "goes crazy," so to speak, at very small distances, causing problems that were long considered unsolvable. These problems manifest themselves in the form of mathematically meaningless quantities, called divergences. Practically, this means that as soon as we start calculating a physical effect, we get an infinite value, like the sum

$$1 + 2 + 3 + 4 + 5 + 6 + 7 + 8 + ...$$

If we calculate a "small" correction to the same effect, we again get infinity, like the value of

$$1^2 + 2^2 + 3^2 + 4^2 + 5^2 + 6^2 + 7^2 + 8^2 + ...$$

And so on. It's clear that a theory with such features is of no use.

Similar problems also arise in the standard model, but they are under control and can be successfully resolved. However, when it comes to gravity, finding a satisfactory solution has not been possible for many decades. And if one tries to remove the divergences in the same way that was successful for the standard model, the resulting theory of quantum gravity becomes infinitely arbitrary. This brings us back to square one: a theory that is entirely unusable, incapable of making predictions at infinitesimally small distances.

In reality, the problem of quantum gravity was not hopeless at all, as everyone thought. In fact, the solution was right there, in front of our eyes, just one step beyond the standard model. There was no need to make assumptions about the nature of spacetime at infinitesimally small scales. There was no need to let our imagination run wild, to rely on our partial and deceptive intuition. All that was needed was to continue the descent into the infinitesimally small along the main path, without deviating, without hesitating. All that was needed was to believe in it. And the answers would have come naturally.

7

At the Heart of the Universe, where Everything Loses its Significance, yet still Governs the Cosmos

As mentioned earlier, quantum mechanics is a disruptive revolution compared to classical physics. In comparison, the standard model is simply a continuation along the path already taken to deepen our understanding a bit further. What can we say about the last step of the descent, quantum gravity? On the one hand, it is just one more step beyond the standard model, along the path paved by quantum field theory. On the other hand, this seemingly small step leads to even more disruptive implications than those encountered with quantum mechanics.

Not only, but it is a remarkable upheaval, compared to what the scientific community had expected for half a century. Indeed, the vast majority of physicists asserted for decades, without hesitation, that it was necessary to abandon quantum field theory in favor of completely new approaches such as holography, string theory, and loop quantum gravity. In reality, the solution to the problem of quantum gravity was within reach, without the need to explore entirely new directions. It was right there, in front of our eyes. If you want to take a life lesson from this book, take this one: the answers to our questions, even the most complex ones, are always in front of us. They look at us and ask, "Why do you continue to ignore me? What must I do for you to notice me?"

In the case of quantum gravity, the discovery I made had been in front of the eyes of all physicists in the world for a century. And even in front of my eyes for many years. But I had never managed to see it. Until at one point, by a strange twist of fate, one of those quantum dice rolls that quantum mechanics offers us abundantly, by looking at things from a different perspective, I finally realized what nature was trying to tell me. And I concluded that, in reality, it had always been there, in front of me.

The theory of quantum fields has brought about the triumph of theoretical high-energy physics, pushing our understanding of nature to the extreme. Its greatest success so far is, as we have already mentioned, the standard model. Saying "high energies" is like saying "small distances" because the energy of a quantum is inversely proportional to its wavelength (its "size"): a quantum has more energy the smaller it is. Therefore, theoretical high-energy physics is the branch of physics that studies phenomena that occur on small scales.

However, it should be noted that quantum field theory operates within a mathematically well-understood formal framework, making it extremely constraining. This means that it does not allow for a great deal of freedom, leaving little room for new ideas. In fact, for decades, no one succeeded in proposing innovative ideas in this field. That's why many people, in desperation, abandoned it early on to propose all sorts of alternatives, like the ones mentioned earlier: string theory, loop quantum gravity, and holography. Finding space for original ideas within quantum field theory was a challenging task. It was like looking for a needle in a haystack, like searching for the one ring that does not hold the chain. What was necessary was to explore the whole framework thoroughly, searching for unnoticed elements, to check whether something crucial had been unintentionally overlooked. This has been my task for twenty-five years: probe quantum field theory from its foundations in order to discover what had never been spotted.

Seven years ago, I completely lost hope of achieving this. The theory was so constraining that whenever you tried to propose a new idea to solve a certain problem X, it would generate an even worse problem Y in another sector. For example, if you tried to eliminate the mathematical divergences caused by the graviton at small distances, you ended up creating absurdities elsewhere in the theory, of the type that I now describe.

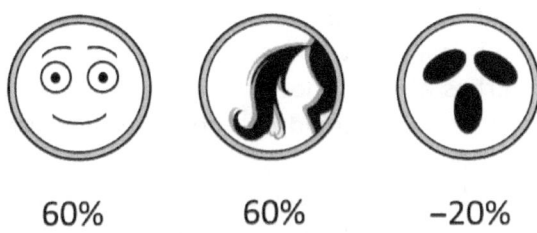

60%   60%   −20%

Fig. 13 | Awkward coins with three sides

The only option that seemed to solve the problem of divergences was to introduce what are called ghost particles, which are particles that lead to physically unacceptable results. We can illustrate them with a simple example. Take a coin. We know that if we toss it, it can land on heads or tails, and both outcomes have a probability of 50%. Now, a theory of the universe that contains ghosts, i.e., ghost particles like those that were used to solve the problem of divergence caused by the graviton, predicts the existence of "three-faced coins" (see Figure 13). These are coins for which each toss, for example, has a 60% probability of landing on heads, a 60% probability of landing on tails, and a -20% probability of landing on the third face, i.e., the ghost result. This makes no sense, as you may note. These are ideas that can be formulated mathematically (add a negative sign in front of a probability? What's the problem?), but they have no physical meaning. Theories that contain ghost particles lack a coherent physical interpretation and must therefore be rejected.

Well, after working so hard in search of the famous "needle in the haystack" and after losing all hope, towards the end of 2016, I began to explore the last corner of quantum field theory that remained to be examined. I had neglected it because I was sure I would find nothing new there, trusting my predecessors, the Nobel laureates who had already explored this field years before and concluded that it could hold no surprises.

Thinking precisely that the great physicists who had preceded me had already clarified everything, I began to study this last domain without much hope, more out of a sense of duty, for the sake of thoroughness. I thought that even if I didn't discover anything really new, I might gain a better understanding of the properties already known, and make some progress by generalizing them. That could be enough to produce an interesting publication.

However, to my great surprise, Pandora's box opened. I realized that the subjects in this research area had not been thoroughly studied at all. And as I delved into my study, the path to quantum gravity emerged before my eyes. It materialized in the idea of a new kind of particle, the fictitious particle I called the fakeon. In short, the fakeon is a particle that can only be virtual but can never become real. It was the only physical entity that could be introduced into quantum field theory without unbalancing it completely, the only possibility that had been overlooked for decades.

In practice, the fictitious particle is a particle that mediates interactions, just as the graviton or the photon mediate gravitational and electromagnetic interactions, respectively. However, unlike the graviton or the photon, which can also be "seen" (especially the photon, as it's precisely what enters our eyes, so we see it literally), the fake particle can only mediate interactions but cannot be directly detected.

Let's take a step back. We've mentioned that the graviton "mediates" gravitational interactions, just as the photon "mediates" electromagnetic interactions. What does this mean?

Take a ball in your hand. If you drop it, the ball falls to the ground. "Obviously," you might say. Let's not jump to conclusions too quickly. How does the ball know there's the ground beneath it? And that the ground isn't beside or above it? Who tells it where to "fall"?

You should know that objects continuously exchange quanta. In the case we're talking about, they exchange gravitons. This is how they know about each other's existence. A continuous flow of gravitons between the ball and the ground ensures that the ball already knows where to go when we drop it (see Figure 14). Similarly, electric charges and magnets exchange a continuous flow of photons, and that's how they attract or repel each other.

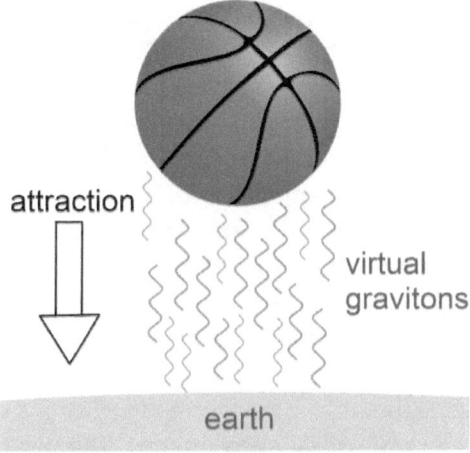

Fig. 14 | Exchange of virtual gravitons between the Earth and the ball

However, the gravitons and photons exchanged by objects are quanta of a somewhat special kind. They are called "virtual" because they are not visible. They travel between one body and another, putting them in communication. This is precisely why they cannot be seen by us, because they never reach our eyes.

It was thought that all quanta could (or should) be of two types, or have two phases: real and virtual. Let's go back to our most important experiment, where a single photon leaves the laser pointer, then passes through the slit, and finally decides where to illuminate the screen (Figure 7). We can say that during the journey between the laser pointer, the slit, and the screen, the photon is virtual. When it illuminates the screen at the place it has chosen, the photon becomes real. And there we see it.

The fakeon, on the other hand, is a quantum that can only be virtual and can never become real. It simply mediates interactions. However, it is not a ghost because it does not lead to absurd results, although it brings about radical changes in our understanding of spacetime at infinitesimally small scales.

How does the fakeon work? How does it solve the problem of quantum gravity? Let's go back to our three-faced coins. In a theory that predicts fake particles, these coins still exist, but now when we toss one, we have a 50% chance of getting heads, a 50% chance of getting tails, and a 0% chance of getting the fake result. In other words, the third face of the coin still exists, but it never shows itself to us. In this way, it doesn't create any absurdity. The fake particle does its job, but it doesn't disrupt the theory, make it meaningless, or cause problems in other sectors. However, as we will see shortly, it has important consequences for understanding spacetime at the infinitesimally small scale, and this will shed even more light on what we are and, most importantly, what we are not.

In the end, we discover that quantum gravity comprises a triplet of bosonic fields to be added to the list of fields from the Standard Model previously encountered. The triplet consists of the graviton g, the fakeon χ, and a third boson φ (see Figure 15). The graviton is the mediator of gravitational interactions and, as such, it is massless, much like the photon γ involved in electromagnetic interactions. The fakeon χ is, in a way, a massive, very heavy copy of the graviton. Being heavy, its effects can only be appreciated at infinitesimal distances (i.e., at high energies, meaning energies greater than its mass). But the two fields alone are not sufficient to explain everything: a third boson, which we call φ, is needed, and we do not yet know its nature for sure, in the sense that it might be real or fake. Only an experiment can answer this question. For now, what we know is that φ is also very heavy, probably as heavy as the fakeon χ.

	Fermions		
Quarks	u	c	t
	d	s	b
Leptons	e	μ	τ
	ν_e	ν_μ	ν_τ

Bosons	
γ	H
$W^\pm$	g
Z^0	φ
g	χ

Fig. 15 | The fields of the Standard Model and quantum gravity could explain all of reality

These are details, so to speak, but they tell us once again, if needed, that the theory is so constraining that it defines its own content. For example, nothing would work if we tried to eliminate this third boson φ. Ultimately, the graviton triplet, added to the fields of the Standard Model, could explain all the reality we know.

Someone might object, "Okay, we get it: your idea is not as arbitrary and artificial as the others. But you still need to explain why it should be the right solution." Clearly, nature has the final say, as always. The validity of a theory can only be confirmed through experimental evidence. It might take some time in this field, but perhaps not as much as one might think. That said, let's delve again for a moment into the historical trajectory of the quantum revolution. Unlike all alternative approaches, such as string theory, loop quantum gravity, and holography, which deviated onto uncertain new paths, the theory of quantum gravity arising from the idea of the fakeon particle is entirely in line with the path that led to the Standard Model. The fakeon is just the last mile of this fruitful road, a small step forward from the theory that has met with the most success in human history. So much so that we can incorporate quantum gravity into the Standard Model without any difficulty. This allows us to build a final theory that explains the four interactions of nature. And its key formula still fits into a single line.

The resulting theory is highly predictive, perhaps one of the most predictive theories of all time, as with few free parameters, it can describe all scales, from astronomical distances to infinitesimal distances. As demonstrated with Marco Piva, a young doctoral student who worked with me on fakeons and quantum gravity, physical quantities can be calculated with an effort very similar to that required for similar calculations in the Standard Model (with some inevitable technical complications). Furthermore, the first quantum corrections we have found already show signs of interpretive upheavals that are reshaping our understanding of the universe. We will discuss this soon.

In the end, the theory is testable and falsifiable, at least in principle. But in practice?

A remarkable property of the idea of the fakeon particle is its universality. It can be applied even in contexts that do not involve quantum gravity. This means that there could be other fakeons that we have never thought of before. For example, are there fakeons among the fields of the Standard Model, the ones in Figure 11? Some particles, including the Z, W, and Higgs bosons, have never been observed directly, but only through the interactions they mediate. In principle, they could exist only as virtual particles. It is possible to indirectly demonstrate, based on the experimental data already available to us, that this possibility is excluded in all cases except for the Higgs boson. The possibility that the Higgs boson is a fakeon remains open and will be tested by next-generation accelerators. And if the Higgs were indeed a fakeon, my theory of quantum gravity could receive indirect confirmation in the near future.

On the other hand, string theory, loop quantum gravity, and holography are defined rather vaguely. Even today, decades after their introduction in physics, they are in an embryonic stage of development. According to numerous indications, they may remain in that stage forever. In these fields, calculations are too difficult or ill-defined, so very few of them can be performed. Moreover, making predictions from approaches like these is extremely challenging. They allow for such arbitrariness that in the case of contradiction with experimental data, it is always possible to adjust the matter to make what doesn't fit, fit. In other words, these theories are not predictive. In contrast, the theory of quantum gravity based on the fakeon particle is so constrained, like the entire theory of quantum fields, that it is essentially unique.

In essence, with the fakeon, we are not making leaps into the unknown, invoking fanciful or groundless hypotheses from our imagination, but rather, we are simply continuing on the main path that nature has shown us. We are taking just one small step further, the only one allowed, beyond the standard model.

The fakeon is a possibility that no one noticed before. It remained well-hidden for at least fifty years. In hindsight, we can confidently say that the fakeon particle could have been discovered half a century earlier, or even a century earlier, because the necessary knowledge was already available at the time. However, such a possibility was never noticed by anyone. Why? Probably because it was not needed, in the past. To build the Standard Model, the fakeon is not strictly necessary (although the Higgs boson could ultimately be a fakeon, as I mentioned). In contrast, the fakeon is absolutely necessary to construct quantum gravity and make it work.

It often happens that when we don't need something, we don't even notice its existence. So, we don't see the possibilities that are right in front of our eyes because we don't need to use them at that moment. It can be very challenging, after 40 or 50 years, to go back, sift through everything that has been done in the meantime, and try to unearth what has been overlooked. The missing piece could be hiding anywhere, in any of the progress made during the intervening decades.

Now we discuss the implications of the fakeon particle. We are almost at the end of our journey into the infinitesimally small. We started with classical mechanics, then moved to quantum mechanics, made a brief stop to discuss the Standard Model, and then descended into quantum gravity. At this very moment, we are in the deepest abyss of the world. What can we learn from it?

What happens in these abysses, at scales a billion billion times smaller than an atom, where the effects of the fakeon particle become significant, is that the concepts of causality and logical implication lose all meaning. Time ceases to exist as such. All temporal order (past, present, and future) collapses. Cause-and-effect relationships disappear. We can no longer speak of origin and end, premise and consequence. The reason for this is precisely that the fakeon is only a virtual mediator and can never become real.

We come to understand that other pillars of our supposed understanding of nature, suggested by the partial view offered by the environment in which we live, such as the relationship between cause and effect, are nothing more than illusions conveyed by the same macroscopic world that surrounds us. They are mirages triggered by what we learn while living in a context of large relative distances. Indeed, at infinitesimally small scales, there is no causality, no possibility of identifying a "before" and distinguishing it from an "after," no possibility of ordering events temporally, causally, or logically. That's the reality of things down there. The theory itself reveals it to us, because, as I mentioned earlier, the theory emerges from the only possibility that remains open, not from artificial hypotheses popping up from our imagination.

To provide a somewhat more concrete idea of how the loss of causality and temporal order manifests, consider the following. The physical equations of classical mechanics are such that the future is uniquely determined by the present and the past. It's thanks to this that we can put a satellite into orbit: we know in advance what it will do. However, if we try to describe phenomena explained by quantum gravity with analogous equations, it turns out that to predict the future, we need to know not only the present and the past, but also at least a little bit of the future itself. To predict what will happen, say, in a second, we must know in advance what will happen in the time span that elapses between now and a tiny fraction of a second. This tells us that everything that happens between now and that minuscule fraction of a second is entirely out of our control. Not only is it undeterminable, but it's unknowable. It makes no "sense."

Another non-trivial implication of quantum gravity is that there are cases where this new indeterminacy becomes relevant even on a cosmic scale. In the current view, crucial events occurred just after the primordial explosion, the big bang, which shaped the universe into what it is today. However, the duration of these events is exactly the same as the uncertainty we just discussed. In the end, the entire universe could be the product of the "anarchy" predicted by quantum gravity on an infinitesimally small scale, meaning the absolute freedom that prevails there. How this could have given rise to an approximately comprehensible universe, we do not yet know.

So, we discover that, ultimately, the abyss of the world is the receptacle of the complete absence of meaning, the violation of all rules. It's total anarchy, absolute freedom, to the extent that it does not even allow us to order events, neither temporally, causally, nor logically. At the same time, we expect that what happens on the infinitesimally small scale explains what is larger. Or at least that's how it should be. It's what we've learned from our approximate way of seeing things throughout our entire life. Even though, at this point, it would not surprise us if we discovered that the idea that what is smaller determines what is larger is also incorrect. In any case, we must conclude that the entire universe is governed by an abyss where nothing makes sense, no rules apply, everything is allowed, total anarchy reigns. For now, we stop here. Let's just accept this message and treasure it. It's a bit like a tile in a mosaic, or a piece of a puzzle. Maybe one day we will manage to gather enough pieces to guess the underlying picture. Perhaps we will conclude that it doesn't conceal anything accessible to us.

It is becoming clear that it wasn't easy to anticipate the upheaval of spacetime that was about to emerge at such small distances. Once again, we realize that our thinking, shaped by the ordinary world in which we live, does not allow us to develop keen intuition about a profoundly different world. The only possibility was to hope that the theory itself would guide us in the darkness. And that's exactly what happened. The reason is that the theory is highly constrained and offers no room for maneuver. The theory tells us what solution emerges from the idea of the fakeon, and that it is unique. All that remains for us is to study it and derive its consequences, which will also be necessarily unique.

Perhaps, as we descend to even smaller scales, we may reach a point where we must surrender. Then, it will be "game over" for us, humans. We will no longer manage to find a correspondence between our world and the world down there. We won't even be able to utter a single sensible word to describe the unknown. But for now, let's content ourselves with what the journey we've made is teaching us.

In hindsight, we can, in fact, assert that the disruption, which was impending as we delved into the realm of quantum gravity, especially the loss of sense of the notion of causality, had already been foreshadowed by the Standard Model itself. It's not entirely true that the Standard Model was without surprises. What is true is that the surprises were not so evident, there. Indeed, within the context of the Standard Model, no satisfactory definition of causality was ever found, which is why the notions of cause and effect were gradually put aside and forgotten. The fourth step of the descent, quantum gravity, delivered the final blow, removing all doubt.

At last, we have a message to deliver to those who are eager to know the fundamental principles of the universe. Philosophy has been asking these questions for centuries. Now is the right moment to gather the knowledge offered by physics and provide some answers. First and foremost, we must note that among the fundamental principles of nature, causality is no longer present. In fact, it has been absent for quite some time, before quantum gravity, simply because in the theory of quantum fields, one cannot even formulate the concept satisfactorily.

Certainly, over the years, several proposals were put forward to define causality in quantum field theory, but they were all artificial, deceptive, and difficult to elevate to the status of fundamental principles. A sign, in the end, that the very notion of causality was on the verge of collapse. With quantum gravity, the difficulty became impossibility, as the theory clearly predicts that causality loses its meaning.

So, once again, one might wonder: what are the fundamental principles that emerge from the theory of quantum fields, i.e., from the Standard Model and quantum gravity? They essentially boil down to three, namely:

unitarity
locality
renormalizability

Now we explain what these principles mean, but let's not be surprised if they have little or no intuitive content. In fact, we will once again have the opportunity to confirm how different the microscopic world is from ours and how weak our intuition is.

Unitarity is the requirement that probabilities be positive or zero, but not negative. Essentially, those ghostly particles we mentioned earlier should not exist. Fortunately, one might say. Of course, this requirement seems obvious, but as we have said many times, we can no longer be sure of anything. We can't even assume that the words we use in our language, even those with obvious meaning, survive the journey we are on. In the end, we must be prepared for anything. But fortunately, at least for now, we don't have to give up the notion of probability.

Locality is the requirement that all of physics can be described by point-like interactions, also known as local interactions. To better understand what this means, imagine two particles, a and b, colliding at a point and generating a third particle, *c*, as illustrated in the left drawing of Figure 16. Perhaps the third particle decays immediately, giving rise to two other particles, *d* and *e*, which can be either identical to, or different from, the initial particles *a* and *b*. In that case, the intermediate particle *c*, described by the wavy line in the right drawing of Figure 16, is virtual and mediates an interaction.

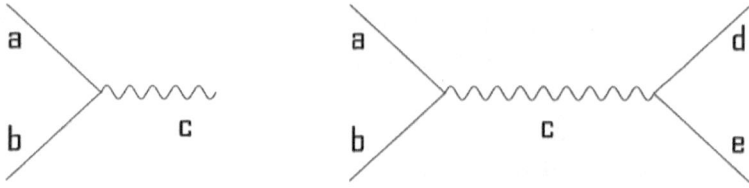

Fig. 16 | Local interactions

These are typical processes studied at CERN in Geneva. In practice, particles are accelerated to speeds close to the speed of light and collide with one another. The collision produces other particles, like *c* in the right drawing, which, however, much like the Higgs boson, only exist for a very short amount of time, meaning they decay immediately after their creation. The products of their decay are detected by particle detectors. Comparing the final particles *d* and *e* with the initial particles *a* and *b* gives us information about the intermediate particles *c*.

This is how we "observed" the Higgs boson and the Z and W bosons. In reality, we never actually saw them, because they don't live long enough; they don't even have time to reach our detectors. We simply observed the decay products *d* and *e*. From these, we inferred that the Higgs, Z and W bosons must exist, or else the decay products would be different or distributed differently.

As mentioned earlier, in the end, the Higgs boson could be a fakeon, in which case we can never see it directly. But we can still see it indirectly, as was done at CERN in 2012. Furthermore, we can even determine whether it's a real particle or a fakeon. In fact, the distribution of the products *d* and *e* predicted by a fake H is slightly different from the distribution predicted by a real H. The definitive answer to the question of whether the

Higgs boson is a real particle or a fakeon will come from experiments planned for the coming years.

Is it true that elementary interactions are local and therefore "simple" in nature? In a sense, yes, in a sense, no, in a sense, maybe. As always, we should not think that our description of reality is reality, especially because we are not there. And we absolutely cannot see beyond the limits imposed by the indeterminacy principle. In particular, we cannot be on the backs of the particles and see what they actually do during their lifetime, whether they are born and die at precise points or in more extended areas. These are all stories that we tell ourselves. They work as long as they work.

And so, what principle is locality, really? It's a formal principle that refers more to our way of formulating the theory than to the actual reality of things. It's the demand that the gaps in our observational capability, i.e., what happens during all the moments when we're not observing reality, can be filled with virtual point-like interactions, which, among other things, are those that we can more easily handle from a mathematical perspective.

In the end, what is virtual, not real, is not described by the misleading hidden variables. It is also no longer described by the outdated wave function of quantum mechanics. In the theory of quantum fields, it is described by an infinite sum of processes like those illustrated in Figure 16.

The third requirement, renormalizability, is the condition that the mathematical divergences we talked about earlier can be removed without leading to an infinitely arbitrary theory. It's also a formal principle, and only refers to the problems we encounter when formulating the theory. You might wonder: could there be a more arid and less intuitive "first principle" than renormalizability?

If the reader is confused, he or she should not worry. Physicists are just as confused. And the deepest meaning of our discourse is precisely this: farewell to intuition, welcome humility. It's the humility of standing before nature like students before their teacher. That is, to wipe the slate clean, eliminate all preconceived ideas, and pay attention to what nature is telling us, in order to learn as much as possible.

Nature has never signed a contract with us. It has never committed to being known or understood by us. And it has never really given us reasons to consider ourselves special in the universe, despite our arrogance that always pushes us to think otherwise. And if we have managed to make progress as much as we have, despite all the difficulties we have encountered along the way, we should really consider ourselves fortunate, rather than getting intoxicated with the idea of our supposed, albeit non-existent, "superiority" over the rest of nature.

8

Discovering the Hidden Meaning of the Universe

In the end, by combining classical mechanics, quantum mechanics, the standard model, and now quantum gravity, we discover that the universe has a very specific direction when it comes to scales of magnitude. What I mean by this is that the world varies with a certain regularity as we transition from large relative distances to small relative distances. At large distances, especially those of astronomical order, but also to some extent our own, determinism prevails, meaning the complete absence of freedom. Events are predetermined there, meaning that the future inevitably follows from the present and the past, from initial conditions. There is no room for vitality or creativity. As we delve deeper down, things change. The metamorphosis starts at the atomic distances, where we encounter quantum mechanics. There, quantum indeterminacy allows the atom to decide for itself whether it should veer to the right or to the left when passing through a slit. We cannot impose anything on it or give it orders. We cannot predict what it will choose to do. The phenomenon is a purely creative one, meaning it produces consequences that do not stem from initial conditions. It yields a result that cannot be foreseen. Then, as we descend to even smaller distances, we discover that the very concept of "determination" ceases to exist as a theoretical possibility because time loses its meaning. There, we can no longer order events or speak of cause and effect.

Of course, in the theory of quantum gravity that I have formulated, there still exists an entity called temporal coordinate, just as there is one called spatial coordinate. However, these are merely mathematical artifacts that at very small distances have no physical meaning akin to what we commonly attribute to them. Once again, we must conclude that our perception, in this case the perception of time and space, is somehow an illusion and an approximation that emerges at our relative scales, the macroscopic distances. But it does not have intrinsic real meaning.

At this stage, we might wonder what truly holds intrinsic meaning in our universe, what still makes sense to say. If words like "origin" and "end" lack fundamental significance, it means they are only rough approximations, useful for understanding ourselves in daily life but much less useful when we want to comprehend the universe. And if that's the case, if it's true that at the fundamental level, nature even rejects notions like cause and effect, premise and consequence, then it means there is no point in asking about the origin of the world, why it exists, what its cause is, and what its purpose is. These questions are just another mystification resulting from a nature that enjoys mocking us.

Let's return to the directionality of the universe that we have discovered as a result of our journey to the infinitesimally small. Gradually, as we descend to smaller and smaller scales, we notice increasing freedom, up to the total anarchy of quantum gravity, where it no longer makes sense to talk about what is a cause and what is an effect, what implies and what follows. Conversely, if we move in the opposite direction, from small to large relative distances, what we observe is the opposite progression, that is to say, an increasing loss of freedom and a gradual shift toward determinism, still life, complete pre-determination, and total predictability. You can refer to Figure 17 regarding this.

We must conclude that the universe is irreversible relative to the scales of magnitude. We can speak of radial irreversibility or radial arrow: freedom increases from large to small distances, up to total anarchy, and decreases from small to large distances, up to total determinism. The universe cannot be radially reversed. In other words, we cannot swap the roles of small distances with those of large distances.

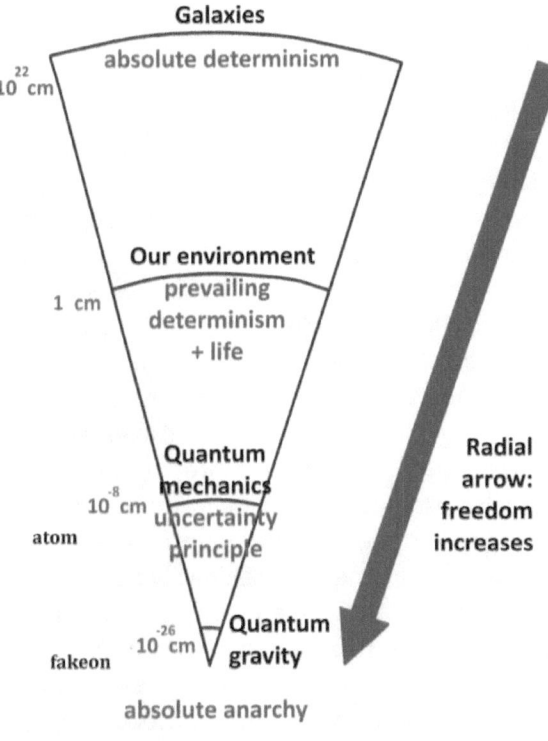

Fig. 17 | The radial arrow

This is the general rule. Does it have exceptions? Is there a countercurrent movement that tries to resist this suppression of freedom, to oppose and extend freedom to larger distances? Yes, there are exceptions. At least one. The most significant countercurrent movement, as we have already mentioned, is life. Life opposes the predominant movement. It fights to find the way, to open a breach and extend the freedom of quantum indeterminacy to ever larger distances. It is an exceptional movement because it occurs in a very limited number of cases overall.

We have just uncovered the hidden secret of the universe, which will explain why we are here, our role, and our mission. We will discuss this in detail soon. But first, let's clarify some points about the concept of irreversibility.

The reader may have heard of another type of irreversibility, temporal irreversibility, meaning that time has a well-defined direction. It flows "forward," so traveling backward in time is impossible. Let's take a short break to compare this type of irreversibility to radial irreversibility.

One field of physics we haven't mentioned, as it's not central to our inquiry, is statistical mechanics, which studies the behavior of systems composed of many atoms, such as gases. Since it's practically impossible to study the evolution of these systems using the equations of motion from the initial conditions (positions and velocities of each atom), statistical mechanics focuses on studying the evolution of average quantities or collective properties, such as pressure, temperature, and volume. "Heat," for example, is nothing more than a way to collectively quantify the motion of atoms. In fact, when we strike a piece of iron with a hammer, we can easily observe that it heats up until it glows red. What we are doing by hammering it is transferring the motion carried out by our arm (i.e., kinetic energy) to the iron atoms. This action forces the atoms to vibrate at a higher frequency around their equilibrium positions. When we approach our hand to touch the piece of iron, our skin receptors send signals to the brain, which the brain interprets as the sensation we call heat.

Temporal irreversibility is generally associated with the law of increasing entropy, the second law of thermodynamics. The entropy law states that in an isolated system, a certain function of the system's state, called entropy, statistically increases over time. Since entropy somehow measures the amount of disorder in the system, the law is often popularized by saying that "disorder always increases with time." A popular version of the entropy law exists. It's the famous Murphy's Law, which says that "if anything can go wrong, it will." There's some truth behind Murphy's Law because if we count the number of ordered configurations of a system and compare it to the number of disordered configurations, we find that the former is very small, while the latter is very large. Assuming that all configurations are equally probable, as is indeed the case under very general assumptions, it's evident that an isolated system (i.e., not "directed" from the outside) is much more likely to evolve toward one of the many disordered configurations rather than one of the rare ordered configurations, which it cannot identify or choose from among the others.

Now, let's explore the analogies and differences between radial irreversibility and temporal irreversibility as described by the entropy law. For the sake of precision, the latter is a statistical law, a law of large numbers, and therefore cannot be genuinely placed on the same level as a fundamental property of nature, such as radial irreversibility. However, in both cases, we're dealing with prevailing behaviors, and it's possible to find exceptions to the rule, meaning minority countercurrent movements. In the case of radial irreversibility, the exception is life, meaning the amplification of the effects of quantum indeterminacy from atomic distances to macroscopic distances. In the case of the entropy law, the exceptions are local fluctuations where entropy decreases instead of increasing, even in an isolated system. But there's a crucial difference between the two types of minority movements.

Indeed, in the case of temporal irreversibility, there's no possibility of amplifying the minority movement that opposes the prevailing movement to make it last longer than a mere fluctuation. However, this possibility exists in the case of quantum indeterminacy, and it might be the most important property of the universe, at least for us, as it leads to life. As explained earlier, amplifying from the small to the large is a monumental endeavor. The mathematical calculation of probabilities leads straight to a binary choice, an either-or situation. Either amplification is facilitated by a multi-step mechanism that dilutes the effort over about ten steps, each amplifying by a factor of about ten, or there's no hope. In the case of radial irreversibility, this mechanism exists and enables the amplification of quantum indeterminacy on a macroscopic scale. By favoring the minority movement that opposes the prevailing movement, it provides it with the possibility to emerge, expand to large distances, and reach otherwise unattainable objectives. However, nature has not provided any similar stepwise mechanism in the case of the entropy law. Therefore, statistical fluctuations remain confined to their natural domain: specific cases, exceptional and extremely rare events. A long-term decrease in entropy can only occur on a timescale much, much longer than the lifetime of the universe.

Often, temporal irreversibility is associated with the direction of time. In reality, what our journey into the infinitesimally small has taught us is that at that scale, time has no meaning and hence no direction because we cannot distinguish the present, the past, and the future. The direction of time is not a fundamental property of nature but a property that makes sense at scales immediately larger than those of the infinitesimally small.

What can we say about the anarchy that reigns in the abysses of the universe? Is there a multi-step mechanism capable of amplifying its effects to produce an impact at much larger distances? Well, no, there isn't. Otherwise, according to the binary choice mentioned earlier, its effects would have already manifested themselves in all possible and imaginable forms and situations. Consequently, this anarchy remains confined down there. We can call it asymptotic anarchy, to emphasize that it can only be reached in the limit of infinitesimally small scales.

This is why time has a very specific direction for us. The direction is not given to it by the entropy law but by the properties of quantum gravity, combined with the absence of a suitable stepwise amplification mechanism to amplify the typical anarchy of infinitesimally small distances. This is what shapes the universe into its current appearance.

We have listed a number of very powerful consequences of the crucial binary choice, the 0/1 bifurcation, to which probability calculus, which governs the amplification from the microscopic to the macroscopic, inevitably leads: either nature provides a ladder, a stepwise mechanism with the required properties to facilitate the amplification, and then amplification occurs whenever it can, or this ladder doesn't exist, and then amplification never occurs. The only known case where this stepwise mechanism exists is the case of radial irreversibility, and it leads to life. No similar mechanism exists in the case of the entropy law, so its violation is limited to minor fluctuations. Similarly, there's no analogous mechanism in the case of asymptotic anarchy. That's why, on a macroscopic scale, we must resign ourselves to the fact that time has a well-defined direction, and traveling backward in time is impossible.

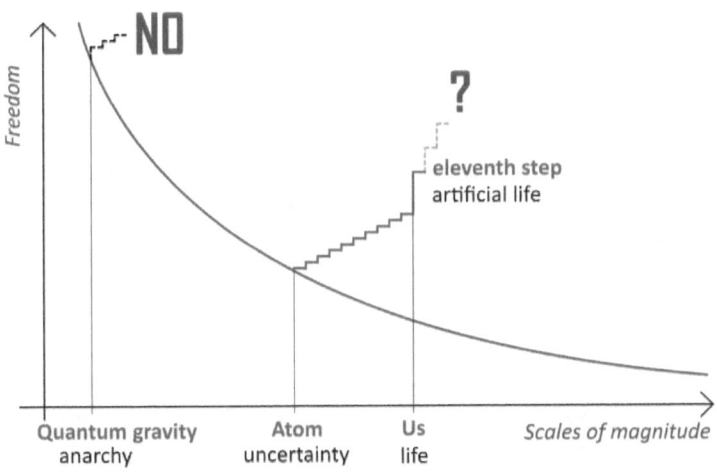

Fig. 18 | There exists a step-by-step amplification mechanism for quantum uncertainty, but no such mechanism exists for asymptotic anarchy

If we want to be precise, we can say that traveling backward in time is possible only... within the limit where it makes no sense or is entirely useless, meaning for very short time intervals, on the order of 10^{-36} seconds (that's one billionth of a billionth of a billionth of a billionth of a second), which is where the purely virtual particle, the fakeon, prevails. In this case, however, our time travel remains confined to the asymptotic anarchy of the infinitesimally small, where we cannot distinguish before from after. Only there, by stretching the argument to the extreme, can we delude ourselves into traveling forward and backward in time at will. But this would be completely pointless, because no event can occur in such a short time, during which it makes no sense to speak of temporal order. Going "back" by so little can't be used to undo anything we've done, because *we haven't done it yet*!

So, we must get used to the idea that time has meaning at our scales, but does not exist at infinitesimal scales. Like many of the concepts we use in our daily lives, it is an "emergent" notion, not a fundamental property of nature. It allows us to understand each other and provide an approximate yet effective description of events within our environment. However, it becomes deceptive when we ponder over first principles, being, and the universe.

Going back is not possible because time is ordered, has a defined direction at all distances greater than those of the infinitesimally small. And because the laws of physics do not allow the amplification of "anarchic" effects specific to those scales, which are the only ones we could rely on to make sense of something like time travel. Nowhere in the universe can anarchic effects be spontaneously amplified over long periods of time.

I used the adverb "spontaneously" on purpose because we are still in the realm of arid probabilistic laws. In reality, a (distant) possibility to bypass this obstacle might exist. It is offered by life itself.

Let's explain this idea better. We have identified two types of "freedom" in nature, on a microscopic scale: quantum indeterminacy, which manifests itself at atomic scales, and asymptotic anarchy, which manifests itself at infinitesimal scales. The former admits a mechanism of spontaneous amplification from the microscopic to the macroscopic, which is life as we know it. The latter does not admit a similar amplification mechanism. In this case, the challenge of achieving amplification rests on our shoulders, as well as on the shoulders of any other intelligent beings inhabiting the universe.

The task is to find a non-spontaneous, artificial way to achieve the goal of amplification. Life itself should take care of the challenge through its most evolved species. Just as in the case of artificial life (which represents the eleventh leap in the amplification of quantum indeterminacy), the leap upward from asymptotic anarchy is heavily disadvantaged by the laws of statistics. Only intelligent beings can overcome the obstacle and accomplish the goal. This allows us to understand the role of intelligence in the universe.

The amplification of quantum indeterminacy has succeeded in making ten upward leaps on its own, spontaneously, without assistance. On the other hand, the eleventh leap, which is the realization of artificial life, cannot occur spontaneously, because it must fill a huge gap compared to the previous ones. Again, it is penalized by the brutal laws of statistics, which make it exceedingly improbable.

Therefore, the eleventh leap, the creation of artificial life, necessarily requires the existence of intelligent beings who take care of it. Only they can achieve this goal after discovering the physical laws and learning how to use them. It is taken for granted that intelligent beings like us already exist on billions of other planets. Perhaps many of them are already working towards this goal, and some may have already achieved it.

On the other hand, the amplification of asymptotic anarchy cannot occur spontaneously because there is no hierarchical mechanism to facilitate it and counter the penalizing laws of statistics. However, just as intelligent beings populating the universe can enable life to make the eleventh leap by creating artificial life, they could also enable the first giant leap (equivalent to eleven, in the absence of the previous ten) of asymptotic anarchy.

It remains to be seen whether the physical laws are designed to allow this. In truth, we cannot say yet because the study of quantum gravity is still in its infancy. Accessible paths for the amplification of asymptotic anarchy (as well as paths to prolong the decrease in entropy over time) may or may not exist. If they exist, they may be convenient or not. It is too early to tell. In the meantime, we can focus on creating artificial life, because the accumulated knowledge so far is sufficient for this purpose. And ultimately, leave the other two challenges to our artificial descendants.

So, we have identified the role of intelligence in the universe. Intelligence is the tool that nature has devised to artificially enable the amplification of microscopic freedom when the laws of statistics make it spontaneously impractical. The role of intelligence is to fill a gap, to remedy a deficiency in nature, namely the absence of tiered amplification mechanisms. The existence in nature of a single tiered mechanism, the one that allows the spontaneous amplification of quantum indeterminacy from atomic distances to natural life and then to intelligence, can compensate for the absence of similar tiered mechanisms in all other cases.

The creation of artificial life is within our reach. And it is within the power of many other intelligent species populating the universe. This goal will be achieved, whether here on Earth or elsewhere, and it will become part of the history of the universe, giving it a new direction.

Thus, if the more or less intelligent living beings scattered throughout the universe find a way to amplify the asymptotic anarchy predicted by quantum gravity, or to violate the law of entropy for extended periods of time, all the better. Such results could open up unimaginable scenarios. Conceivably, these two other types of amplification require forms of intelligence much more powerful than ours, because they aim to reach macroscopic distances from infinitesimal distances in a single leap, without being facilitated by tiered mechanisms.

We, the human species, are an instrument in the hands of nature. Nature uses us to search for artificial ways to allow microscopic freedom to make leaps that are strongly disadvantaged by the brutal laws of numbers, which oppose spontaneous amplification. We are called to play the role of the missing piece. And, like us, nature employs other intelligent species present on other planets throughout the universe. We, like the others, are trials, attempts, throws of the dice. Many attempts are necessary for at least one to succeed.

We have learned that the meaning, mission, purpose of life, the reason life has appeared in the universe, is to resist the suppression of freedom to which the macroscopic world is otherwise condemned, and to seek ways to amplify this freedom and make it more powerful. Our task is to persist in this endeavor, fortifying it in numerous diverse directions, much like those we are exploring in this book, breaking all barriers, all patterns, all constraints, up to overcoming the physical laws that bind us. And to use intelligence to achieve unprecedented leaps.

We might wonder: will we, as human beings, be up to the task? In principle, yes, because the knowledge accumulated so far is sufficient to create artificial life. But that doesn't necessarily mean we will actually do it. The point is that human beings are essentially selfish. The creation of artificial life is an enterprise that could surpass us as a species, because it requires an altruism that may not come naturally to us. Throughout history and progress, we have mainly thought of ourselves, sought ways to improve our living conditions, expanded human presence in the world, disproportionately increased the population, extended individual lifespans, eradicated diseases, or cured as many as possible. We have eliminated or contained wars, causes of death, sources of conflict. All of this, even at the cost of depleting the planet's resources, endangering many living species, and causing their extinction.

Succeeding in creating artificial life forms is evidently an altruistic endeavor, as these living beings could achieve outcomes beyond our reach. It implies that we must eventually pass the torch of quantum indeterminacy amplification to these life forms. We would no longer be the central figures in this grand adventure called life. It's akin to accepting a supporting role and relinquishing the lead. Forever!

In short, the human species will have to cede the scepter to other beings, created by itself. They will continue the exploration with capabilities and tools more powerful than ours. They will travel through space on our behalf, colonize the universe, bring artificial life forms (and perhaps natural life forms as well) to many other planets, and then continue the process of amplifying quantum indeterminacy to even larger scales.

Since the creation of artificial life is an altruistic project, one might ask: why should we ever do such a thing, even if we can? We may not benefit from it. According to the scenario outlined in this book, one day we could become slaves to the beings we create ourselves, the q-droids. Or we could be placed in their zoos, where we would be observed and controlled from morning to night, treated as we treat other animals today.

When those ten million surviving humans will live confined in Australia, it is highly likely that the q-droids will completely control their existence, stripping it of all meaning and purpose. They will decide which people will be sterilized and which will not, which men will have children with which women. Exactly as we do with the animals we keep in cages. Humans will be deprived of all freedom and stimuli. And from time to time, the q-droids will visit Australia as tourists to get to know "those who created them."

Fundamentally, our predecessors in evolution experienced similar fates. They did not know what to expect. They were not aware that they would produce a being that would later put the entire planet in check and cause the extinction of many species and the complete enslavement of many others.

The difference is that, in our case, the creation of artificial life should occur consciously. The eleventh leap of amplification cannot occur otherwise, as it is a much more demanding and difficult leap to achieve than the previous ten. Therefore, it is impossible for it to happen spontaneously in nature without the intervention of an intelligent being.

Human beings will face this dilemma. A decision will be made, "consciously" or not. The decision will not be "willed," as we know. It will be generated by quantum phenomena occurring in the brains of human beings. However, one thing is clear, as it has become very evident during our journey: what we have discovered is the physical meaning of life and its mission.

Now we know what life is. We have identified the role of intelligence. We know where we come from, we know our place in the universe, and we also know where we are going, that is, what we are called to do.

Life is the amplification of quantum indeterminacy from microscopic scales to larger and larger scales. Natural life has done its part and succeeded in reaching our scales. Using our intelligence, we are called to ensure that the amplification makes the eleventh leap, which will allow it to continue its journey and perhaps reach arbitrarily large distances.

Now that we know what we are doing in the universe, why we are here, it will be up to us (that is, the dice rolls that occur in our brains) to decide whether we proceed with our mission or not. But life will continue along its path independently of us. And considering that the universe harbors billions upon billions of inhabited planets, we must be aware that even if we do not make the eleventh leap, someone else, on one of these planets, will do it in our place. Possibly someone is already thinking about it. Maybe someone has already done it. And perhaps one day, superior artificial beings will arrive on Earth from somewhere in the cosmos. Then, the destiny of life on Earth will be fulfilled nonetheless. But likely we won't be here to witness it.

9

The Correspondence between the Large and the Small

In the early days of quantum mechanics, faced with the challenges posed by the interpretation of new experimental data that revealed a microscopic reality that was very different from the macroscopic one, some researchers, notably Niels Bohr, attempted to describe this diversity by means of a "principle of correspondence" to identify potential relationships between the small and the large. Studying these correspondences is useful for us to contemplate certain aspects of the journey we have undertaken and better understand the message that has emerged from it.

Let us consider the atom. Before quantum mechanics, various models were put forward to describe it. The two main ones are due to Thompson and Rutherford. Thompson imagined the atom as a "panettone" (Christmas Italian cake), that is, a positively charged sphere inside which smaller, negatively charged spheres (the electrons) are distributed at fixed positions (like raisins in a panettone). Thompson's model was abandoned quite early, as it was inadequate in several respects. The main flaw was that this kind of atom is "full," while in nature, the atom is essentially empty.

Rutherford proposed a planetary atom model, where the positive charge is concentrated in a fixed central nucleus, around which electrons orbit, much like planets orbit around a star. Rutherford's model is also inadequate and flawed. We have learned that we cannot even say that the atom exists when we are not observing it. In particular, after passing it through a slit, we cannot say that it exists during its journey to the detector. Similar conclusions apply to electrons inside the atom. So, it doesn't make much sense to describe the atom as a set of beads orbiting one another.

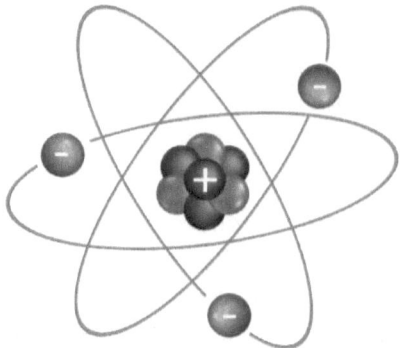

Fig. 19 | Rutherford's planetary model of the atom

This brief historical retrospective confirms once again, if it were needed, that our intuition based on phenomena observed at macroscopic scales appears very naive when we explore smaller scales, and has little chance of getting it right. It is clear that to tackle the problems posed by quantum mechanics, something truly new is required. However, if this novelty did not have at least some type of connection, or "correspondence", with what we are accustomed to, what we can describe using our language, we would not be able to say much.

So, the focus was on searching for relationships, particularly correspondences, between the reality around us and a reality as different as the microscopic one. It was understood that these correspondences, if they existed, might be quite vague, potentially involving ambiguous and approximate concepts.

The solution found is as follows, in the case of quantum mechanics. We can continue to describe microscopic reality using the same quantities we commonly use, such as position, velocity, energy, etc. However, these concepts must be deeply reinterpreted, because they acquire an entirely new physical meaning, and a different relationship with the measurements we make. The atom has nothing to do with a planetary model, because electrons are not beads. They are not localizable, and as long as we do not observe them, they exist nowhere. It does not even make sense to speak of the trajectory of an atom passing through a slit and heading towards a screen. However, quite surprisingly, the basic formula that describes the atom is practically the same as the one we would use to describe Rutherford's planetary model. With the caveat that this formula must be radically reinterpreted, along with each of its components (position, velocity, energy). The principle of correspondence is essentially the guide for this reinterpretation. As imperfect as the whole procedure may be, we have to settle for it, because we can hardly do better.

In the end, quantum phenomena are described by formulas similar to those that describe classical phenomena, but the meanings and roles of different components must be rethought and understood in a new way. The correct reinterpretation historically emerged from the confrontation between different hypotheses and experimental data, after a multitude of trials and errors, failures and successful attempts. Among the failures, we can mention hidden variables, which were supposed to explain the quantum world as a classical world under false pretenses.

The journey to the infinitesimally small, from our scales to atomic distances, and then from atomic distances to those of quantum gravity, can be described as the evolution of this reinterpretation, i.e., the discovery and refinement of the rules governing the correspondence between the large and the small. The fakeon, that is, the possibility to quantize certain fields in a new way, as fictitious particles that are only virtual and never real, represents the latest frontier of this reinterpretation.

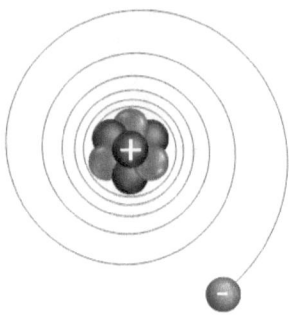

Fig. 20 | In the planetary model, the electron loses energy and falls right away on the nucleus

The reinterpretation in quantum physics is not a simple adjustment but a profound revolution, to the point that the physical results differ completely from those of classical physics. In fact, it was well known from the beginning that Rutherford's atomic model was inadequate, because it predicted that the universe would collapse in on itself and disappear in a fraction of a second. This is because the equations of the electromagnetic field dictate that a charged particle in rotational motion, such as an electron orbiting the nucleus in Rutherford's model, emits energy in the form of radiation. This would gradually slow down the electron, causing it to move along a spiral trajectory and rapidly fall into the nucleus (see Figure 20). The time required for this fall can be calculated easily and turns out to be a tiny fraction of a second. Since the nucleus is a hundred thousand times smaller than the atom, in much less time than it takes to blink, poof!, the entire universe would collapse and basically disappear.

To be precise, even planets orbiting the Sun emit energy in the form of gravitational waves, but in this case, the amount of energy they lose is so minimal that the solar system is not in any danger for the remainder of its existence. All of this is to say how risky it is to transfer models that work in one context to a different context. However, if we think about it carefully, our minds can hardly do better than this copy-paste of ideas. Therefore, one possibility, perhaps the only one we have, for understanding the microscopic world is to begin with gross assumptions, such as Rutherford's atomic model, and then refine them as we progress. This process may require us to reinterpret all the concepts and terminology we employ, with the awareness that there are no guarantees of getting anywhere.

As mentioned earlier, the results of the reinterpretation are so different from those of classical physics that electrons do not rotate at all. In fact, they are not even localized, and it makes no sense to talk about their position. The only thing we can say is that they occupy certain "energy levels," somewhat like clouds, each of which implies a certain average distance from the nucleus. What matters is that the minimum average distance is not zero, so the electron, whatever it is or is not, cannot "fall" onto the atom, and thus the universe never "deflates."

Well, everything around us can only exist because quantum mechanics prevents electrons from falling onto the nuclei of atoms. As much as quantum mechanics may disturb us, this disruption is welcome, because if we were to switch it off even for a fraction of a second, the entire universe would vanish.

Despite this, the atom continues to be misrepresented in many textbooks in the form of the planetary model. The reason for this is that the atom, for what it truly is, cannot be represented at all. If we truly want to imagine it in some way, the best we can do is to surround the nucleus with a cloud of variable density from point to point. For example, in the hydrogen atom, which is the simplest one, because its nucleus consists of only one proton, the single electron is "spread out" in a spherically symmetric cloud, with higher density within a certain distance from the center (see Figure 21). But it is incorrect to say that the electron is "scattered" in this cloud, because, as we know, we cannot even say that the electron exists when we do not observe it. The cloud describes the probability of finding the electron at each point if we were to search for it. In other words, the electron does not exist (it is virtual) until we observe it, so it is nowhere; if we were to reveal it, it would manifest itself at a point in this cloud, with a probability equal to the density of the cloud at that point.

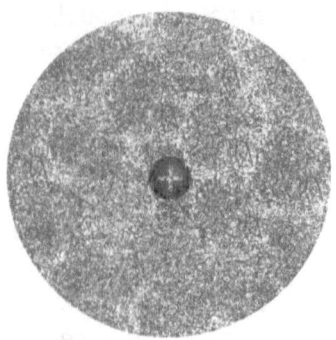

Fig. 21 | Hydrogen atom

The cloud is somewhat like the interference pattern produced by our laser pointer when the beam passes through the slit (Figures 5 and 6). If the laser emits a single photon (Figure 7), the interference pattern represents the probability of the photon ending up in different areas on the screen. The higher the brightness of the area, the greater the probability that the photon will appear in that area.

Despite the profound difference between quantum mechanics and classical mechanics, it is remarkable that we can use the same classical formulas and the same basic quantities. The entire quantum revolution is thus encoded in their reinterpretation, in the correspondence between the small and the large, no matter how vague and imprecise it may be.

When we explore the macroscopic universe, we use our language with its traditional meanings. However, when we approach the microscopic world, we must use the same language, even though we know it is imperfect, and the concepts we use may have entirely different meanings. We must hope that there is some form of correspondence between these two worlds. But this correspondence likely becomes weaker and weaker as we explore smaller and smaller scales of magnitude, eventually disappearing completely, and hindering us from achieving a thorough understanding.

The fact that we cannot describe the microscopic world with the thought and language forged in us by the macroscopic world has not knocked us out yet. However, it has forced us to reinterpret everything, including the very meaning of existence, the nature of observation, and the atom. So far, we have managed to get by in some respects. In other areas, we have begun to encounter difficulties that surpass us.

Descending to the next level, from quantum mechanics to the standard model, the correspondence has already proven to be much weaker than described so far. There, a real connection with the phenomena around us, a true correspondence, no longer exists, replaced by the three arid and unintuitive principles we encountered some time ago: unitarity, locality, and renormalizability. We have said several times that the basic formula of the standard model fits in one line. And the same goes for the formula of quantum gravity. The point is that the reinterpretation of these two lines does not fit in just one line...

What guided us to write these two lines and reveal their reinterpretation? What allowed us to find any correspondence in total obscurity? Surprisingly, it was the difficulties themselves, both in the case of the standard model and in the case of quantum gravity, that came to our rescue to overcome the obstacles. Fortunately, they were there! Otherwise, we would never have been able to say anything.

We mentioned some time ago that if we add the graviton to the standard model in a naive way, it "goes haywire" at infinitesimally small distances, creating huge mathematical problems long thought to be unsolvable: the divergences. Well, the Higgs boson, quarks, Z and W bosons, the photon, also generate similar problems, although much less severe than those of the graviton. What is important to emphasize here is that it is precisely these problems that have guided us toward the correct theory, first the standard model, and then quantum gravity. It is these difficulties that have constrained the theory to tie our hands. Without them, we could have said anything and its opposite, and we would never have had the opportunity to predict anything. Thanks to these problems, on the other hand, the answer is essentially unique!

In fact, there are very few options in which the mathematical divergences disappear, more or less miraculously. All alternatives must be discarded. The elimination of divergences narrows the field down to very few theories. In the case of quantum gravity, there is only one! That's why the fakeon emerges as the only possible solution.

We have learned another life lesson: it is the difficulties themselves that allow us to overcome them. We just need to understand how to exploit them.

The world around us, which we can describe as classical, where quantum phenomena are suppressed, is a rough approximation of reality. It gives us the illusion of determinism, while reality, intimately, is anything but deterministic. Our task of understanding nature is to look for anchor points to describe the microscopic reality from its rough macroscopic approximation, its rough copy. The mental process of quantization starts from a wrong description and works its way up, with the hope of finding the correct description. It is clearly a countercurrent approach that comes with many risks of making mistakes.

In essence, the descent to the infinitesimally small requires us to grasp the truth as a reinterpretation of an error, to reveal the essence of nature from its gross approximation, which is what surrounds us. Not only that, but we must hope that there is indeed some correspondence between the two realities, which is by no means guaranteed. Clearly, candidates for the correct description of reality can be infinite, and they all lead to the same approximate description on a large scale. It is this infinite arbitrariness that constitutes the greatest obstacle to making progress in the exploration of microscopic obscurity. That's why the mathematical difficulties we've talked about are a true blessing. They are our only light in the darkness.

We must explain a creative nature from the limit where nature is not creative at all. It's like trying to understand life from death, deducing the truth from its erroneous, approximate, and rough version, in which most key concepts are lost or distorted, replaced by fallacious concepts like determinism. This makes the process of understanding nature full of unprecedented obstacles. On the one hand, there are obvious experimental difficulties, which involve building expensive equipment and mobilizing many people to search for, as in the case of the Higgs boson, the confirmation of a prediction made fifty years earlier. In addition to that, there are much deeper difficulties inherent in the very process of knowledge. We face enormous conceptual problems that shake our own language and thinking.

It is a very fortunate circumstance that, at scales smaller than those of atoms, a kind of correspondence principle (based on unitarity, locality, and renormalizability) continues to make sense and find success, thanks to the problems of mathematical divergences. This also explains why abandoning the mainstream path to embrace gratuitous proposals like strings, loops, or holograms has no chance of leading to results whatsoever: in those domains, arbitrariness knows no bounds, and possibilities remain infinite.

10

Our Limits

We must confront our limitations. We have dimensions. We are aggregates of atoms trying to understand the world on scales much smaller than the smallest ingredient they are made of. To "understand", we must explore all the paths available to us in search of correspondences between what we do not yet know and what we already know. Probably, this correspondence will become weaker and weaker as we progress to smaller and smaller scales. Fortunately, we have managed to understand something about what happens at atomic distances, and then even further, down to the typical scales of quantum gravity. So, let's continue as far as we can, aware that at a certain point, nature might say, "Dear human being, I'm sorry, but your journey has reached its destination. Beyond this limit, you cannot go."

In reality, nature has already been telling us something similar for decades. We have talked about the great successes of physics, such as the standard model or quantum gravity itself, but we have not talked about the failures, that is, all the situations where we had to stop in the face of insurmountable difficulties. In many cases, we had to deal with problems so difficult that we had to resign ourselves. Forever?

As one would expect, quantum field theory requires highly sophisticated mathematics. It is important to note that the mathematics needed to study the largest sector of quantum field theory (known as the "nonperturbative sector," where we cannot use the main tools of the subsequent approximation) is so advanced that we have not yet managed to develop it. This sector has been explored very little in almost a century, because it is beyond our reach.

We can always hope that, if we wait long enough, we will learn more and make as much progress as we wish. We can hope that a problem that is difficult today will become less challenging in the future and eventually become solvable. We can believe that the progress of human knowledge knows no bounds. However, reality might be very different from this.

Of course, in all sciences, we are daily confronted with challenging problems. What scientists usually do is look for the easiest problems, focus on them, and hope that the future will bring solutions to the more difficult ones. Unfortunately, today, the problems that remain open in high-energy physics are among the most difficult ever, which explains why progress is slow in many areas.

If we want to understand who we are, what role we play in the universe, if we want to estimate our capabilities, it is not enough to report about our successes. We must necessarily account for our failures, discuss our limits, reflect on what we have never succeeded in achieving. We must understand where we had to stop and ask ourselves if the halt can be permanent.

For example, in the theory of quantum fields, mathematical progress has been virtually blocked for about a century. This means that we must use the same mathematical tools that were available to us a century ago. If the tools do not advance, the results we obtain can be significantly limited as a result.

To explain this point in more detail, let us briefly recall mathematical concepts that many readers may have heard of, even those who have pursued technical or humanistic studies, because they are part of the curricula of many high schools, such as calculus: limits, derivatives, integrals. In particular, the integral is a method that allows, under certain assumptions, to define the sum of infinitesimally small contributions, and obtain a finite result. In simpler terms, in many cases, it may be necessary to express a physical quantity as the sum of very many very small contributions. Under certain assumptions, this sum can lead to a finite result (i.e., a value that is neither infinite nor infinitesimal). The tools of calculus are very valuable. They have allowed us to make enormous progress and simplify an impressive amount of reasoning.

The theory of quantum fields could benefit from another leap forward, which could be achieved by formulating a more sophisticated notion of integral, called the functional integral. Mathematicians have been trying to formulate this concept for about a century, but the results obtained so far are partial and unsatisfactory. For now, we have managed to find a number of ways to somewhat bypass the obstacle, but these tricks might not have existed. And similar tricks might not exist in the future. We know that some correspondence between our scales of magnitude and the atomic ones has survived, as well as a correspondence between our scales and those of the standard model and quantum gravity. But it was not at all obvious that this would happen. We could have been forced to stop earlier. Well, the reality that we are beginning to grasp is that we may be forced to stop now. Perhaps we need to start wondering if we have reached the limit of our capabilities.

Let's briefly discuss this functional integral. We know that when a photon or an atom is launched through the slit, it chooses its own path. In the journey between the slit and the screen, it is not possible to talk about a trajectory or even the existence of the atom. We must fill this unknown area with some mathematical artifice that allows us to connect the launch to the final results, meaning the launch to a set of options and probabilities. And this mathematical artifice must be sophisticated enough, because any simplification (such as hidden variables, or the assumption that the atom "exists" even when it is "traveling" and not observed) immediately leads to contradictions with the experiments themselves.

One way out is to imagine that the atom takes all possible paths simultaneously (see figure 22), with a universal distribution of "probability." If it cannot follow a single trajectory, why not think that it follows all of them at once? This idea can only work if we manage to perform a sum of contributions from the trajectories. Trajectories are described by functions, so this hypothetical sum is called a functional integral.

Mathematicians have indeed managed to implement such a program in a very particular set of cases, where it yields correct results. However, they have never succeeded in going further. We cannot even think of doing something similar, for example, in the case of the standard model.

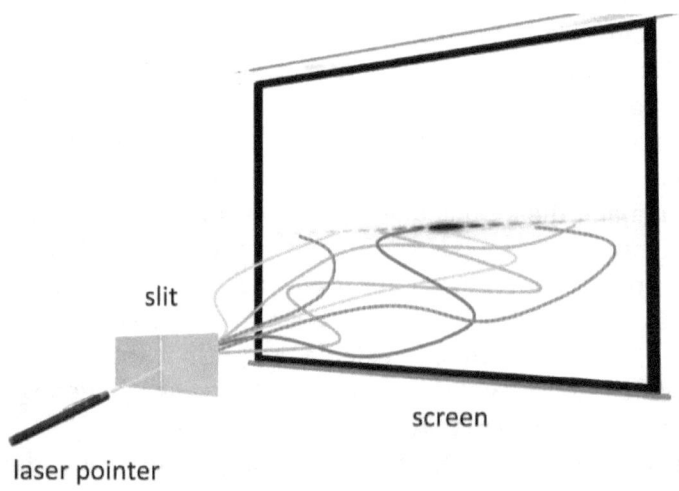

Fig. 22 | The functional integral envisions that photons take all conceivable paths

A shortcut used to bypass the problem has been to proceed through successive approximations. This procedure, called "perturbative expansion," involves considering one interaction at a time. It has led to important results, but has also constrained us to study elementary physical processes, which are the ones we referred to as "local" a while ago, such as those studied at CERN. Complex phenomena involving a large number of quanta simultaneously are beyond our reach. They belong to the "nonperturbative" sector of quantum field theory. There, insurmountable obstacles appear, giving us the impression of having reached the limits of our intellectual capacities.

So, what can we do? What we can do is explore all the possibilities available to us and hope. The point is that we can only try paths accessible to our capabilities. And what if the answer lies in one of the paths that are inaccessible to us?

To better understand our limitations, take a moment to reflect on the true nature of our intelligence. Almost every day, human beings invent arguments to consider themselves superior to the rest of nature. Among these arguments, of course, is the fact that we are intelligent or conscious. On the other hand, those who observe the world with the eyes of science, impartially and non-anthropocentrically, can only conclude that there is no trace of intelligence in the fundamental physical laws of nature. In other words, intelligence is one of the many emergent phenomena. With a significant role, certainly. For example, we have discovered that intelligence is the tool invented by nature to achieve its own purpose, i.e., to allow quantum indeterminacy to make very unfavorable leaps in the amplification process towards larger and larger scales. In particular, it offers the possibility of making the eleventh leap, which is the creation of artificial life. We might think that intelligence is a very special gift that nature has given us in honor of our superiority in the universe. However, that is not the case, to the point that most likely nature has made the same "gift" to billions of billions of other living beings in the universe.

What does "understanding" mean? Is there really a possibility of "understanding" what surrounds us? Or is it just an illusion? When, for example, we encounter a word we do not know, we can look up its meaning in a dictionary. The definition we find allows us to relate the new word to the meaning of already-known words. Therefore, "understanding" essentially means bringing something unknown back to something known, i.e., establishing a correspondence between the two. This is essentially what the process of learning consists of.

There was a previous moment when even the words we know now were unknown to us. Since we didn't know what they meant, we proceeded in the same way, consulting a dictionary or asking others for explanations. By doing so, we brought them back to the meaning of words belonging to an even narrower set, those that were known to us before. If we push the argument to the extreme, going back in time, we arrive at the moment when we were a month, a week, or a day old. At that time, we didn't know the meaning of any word. We simply perceived sounds emitted by the people around us. Gradually, we realized that we could emit sounds ourselves, and that we could emit them in a controlled or controllable manner. We began to reproduce some of the sounds we perceived. We learned that we could associate the emission of these sounds, which we began to call words, with consequences, either positive or negative for us. We started using the words to our advantage. Later, we learned to master this type of action, composing a limited number of sentences. From there, we learned new words and began composing increasingly complex sentences. And then more and more.

These observations reveal that "understanding" reality, in the end, is nothing more than "getting accustomed" to reality. We have accumulated a range of knowledge, realized that using certain words, uttering certain sequences of sounds, elicited reactions around us. We classified those reactions as positive or negative and began using that type of knowledge to our advantage.

It is evident that all of this significantly diminishes the value of what we call intelligence, to which, without realizing it, we have an inclination to attribute a sense far superior to its own meaning, a sense that we assume exists without ever questioning what it truly is, because the moment we ask that question, we inevitably arrive at the conclusion just reached.

We have a strong tendency to take mental shortcuts, imbuing words with meanings they do not truly possess. And often these meanings are unclear or ambiguous. But as soon as we realize that intelligence is nothing more than one of many ways to "get accustomed" to actions, consequences, and reactions, and that understanding new things is just a way of getting used to them, we conclude that when we understand, when we explore the universe, we are doing nothing more than playing the game of chance. We try all the paths available to us, leaving it to fate. If we are lucky, we identify the path we need. Otherwise, we continue searching. This is essentially the same method used, in a different way, by insects or bacteria when they generate millions of individuals, 99.9% of which are destined to perish quickly. Yet, with this method, they manage to discover, through adaptation, how the external environment is made.

Ultimately, in nature, there is no other method than the method of throwing dice, animated by quantum indeterminacy. It is the only unmoved mover. It changes the course of events, causing without being caused, determining without being determined. Not only does it suffice to explain everything that exists, including life, but it is also everything that exists, everything that *animates*.

And I have no difficulty admitting that my discovery of quantum gravity is the result of this same method: exploring all the paths available to me. Even if, unfortunately, I had left the right one for last. You could say that this is the stupidest method that exists. Enough with "understanding"! Enough with "intelligence"! It's anything but that! But it's also the only method that truly exists in nature, and we delude ourselves if we believe that there are others, perhaps more powerful.

When we are faced with the challenge of exploring the unknown and adopt a method like this, that is, exploring all available paths, we implicitly admit that there is no real possibility of "understanding the world." Assigning obscure meanings to the words "intelligence" or "consciousness" that they ultimately do not possess is often a source of error because it precisely prevents us from seeing what is in front of our eyes, as in the case of the fakeon. Why did it remain there, unnoticed, unknown to all for a hundred years? Precisely because everyone thought they had "understood," by attributing a meaning to the word that it did not have, which led them to dismiss several paths right away, with a certain superficiality.

In truth, pursuing all available paths is impossible when taken literally, as we do not have infinite time. We must economize. And that's where the trap lies. I was lucky, for example. What I did to economize was to focus on "all available paths within the theory of quantum fields." Thus, I started from the standard model and tried to take that famous extra step that remained to be taken. I discarded all alternative paths, that is to say, those that did not start from the standard model, but advocated gratuitous and fanciful assumptions about the nature of the infinitesimally small, like string theory, loop quantum gravity, and holography. That's what saved me. Operating this selection may not have seemed so difficult in hindsight. But when you consider the number of people who could have done what I did and didn't, the number of researchers who were misled by a series of mental obstacles and taboos, and the fact that I myself was deceived for many years, you may realize that our psyche is "rowing against the current" without us even realizing it.

Perhaps some of you have heard of the "infinite monkey theorem." Probability theory tells us that if we place a monkey in front of a keyboard and let it randomly press keys, it will eventually compose the Divine Comedy or any Shakespearean tragedy, provided we wait long enough. However, this "long enough" is not within our reach, because it is much longer than the lifespan of the universe. The "monkey method" is not effective, because even though it involves "exploring all available paths," it is completely blind in the sense that each subsequent attempt is independent of the previous ones, meaning the failure of a past attempt is not utilized in future attempts. Such a method can continue indefinitely without ever making progress. And it is a powerless method because even if the correct solution to a problem emerges, no one will be able to recognize it as the right solution among the many incorrect ones, because it lacks the support of experimental evidence and is not the final step along a path already crowned with success.

Making gratuitous proposals out of nothing, as has been done for forty years in high-energy theoretical physics with string theory, loop quantum gravity, and holography, is a good illustration of the infinite monkey theorem. If nothing happens to shake the scientific community, physicists can continue in this manner for eternity without achieving significant progress. We cannot wake up in the morning, propose the first thing that comes to mind, and hope it's correct. The reason is that our intuition is shaped by an environment that is so profoundly different from what we want to explore that we struggle to use the fundamental words of our language.

Exploring mental constructs is certainly a good exercise to train the mind, to be prepared for the right moment when we can finally confront them with experimental data. However, it is not a scientific method, because it is not predictive. Can we say that Democritus discovered the atom because he had the idea of it? Obviously not. Without diminishing the importance of Greek philosophy, as a significant part of our culture originated from it, there is a reason why we call it philosophy and not science. Unfortunately, humanity seems to have a strong tendency to abandon the path of rigor and the scientific method whenever experiments become rare or too difficult to carry out. At this point, humans develop faith in propositions that seem to emerge from nowhere, as contemporary theoretical physics demonstrates. When this happens, the risk is that of getting trapped for decades, or even centuries or millennia, without ever raising doubts.

At this stage, we begin to understand that our journey into the infinitesimally small is testing our intrinsic mental qualities and capabilities. It probably pushes us to confront the inherent limits that hinder our understanding, limits that perhaps only artificial life forms can overcome. All of this should encourage us to be cautious when inventing excuses to place ourselves at the center of the universe, or consider ourselves the pinnacle of evolution. But if that's not enough, there's more.

11

The Threat of Involution

The search for some kind of correspondence between the world around us and the microscopic world is indeed all we can do to "understand" the realm of the infinitesimally small. Every correspondence we find allows us to take a small step forward in this "getting used to" the phenomena that occur at small distances, which is precisely the only true meaning of the word "understand." When we descend into this realm, we struggle to get accustomed to something that is in no way familiar and usual. We need a substantial amount of experiments, data, and confirmations to achieve this goal. The quantity of information we can gather must be abundant enough compared to our brain's needs to allow it to process the data and start getting accustomed to it.

The difficulty we face in high-energy physics today is precisely that we may no longer be able to get accustomed to the world of the infinitesimally small, because conducting experiments becomes increasingly costly. It requires more and more strenuous efforts and longer periods of time. Remember that it took nearly fifty years to confirm the existence of the Higgs boson. If the capacity to "get used to" what is currently unknown diminishes, the likelihood of understanding also diminishes. Quantum gravity was a fortunate case, so to speak, because it was the last mile along an already successful path. Even if experiments on quantum gravity are currently unfeasible, a sufficient amount of data was already available, indirectly, thanks to the standard model. My bet was precisely this: that quantum gravity was "the last link of the chain that contained the standard model." Fortunately, quantum field theory provided enough constraints for me to get used to it quite well. This allowed me to understand what the next step was, which was the fakeon, to transition directly from the standard model to quantum gravity.

Other proposals on the market, such as string theory, loop quantum gravity, and holography, deviated from the main path, instead of continuing to delve deeper. Their weakness was that they had no connections to nature or the standard model. Their proponents relied on the possibility of exploiting a pseudoscientific version of this "getting used to" reality, mainly consisting of indoctrination. They thought that by exerting strict control over the distribution of funding and new hires in universities and research centers, and by getting people used to always hearing the same version, they could render previous knowledge, i.e., the standard model and quantum field theory, obsolete. And gradually reprogram research by acclimating people to artificial pseudo-knowledge.

This led to a plethora of serious problems, the consequences of which the scientific community has to deal with still today. In almost fifty years, there has been little significant progress. The void has been filled with propaganda and pseudoscience. In fact, two or three generations have been skipped. The previously accumulated knowledge has been at risk of being lost forever to the extravagances of those who believed they could reset it on entirely erroneous foundations.

Let's go back to the discovery of the Higgs boson, achieved

at CERN in 2012. As many readers may remember, the news was reported by the media worldwide and reached the general public. However, what the readers may not know, because no one has ever told them, is that the experiment conducted to find the Higgs boson, known as the LHC (Large Hadron Collider), was not primarily aimed at this goal, but at revealing supersymmetry. The discovery of the Higgs boson was the minimal threshold to reach to avoid considering the whole experiment a complete failure.

What is supersymmetry, which rightly no one talks about anymore since then? It is the hypothesis that there is a sort of duality between bosons and fermions, meaning that bosons and fermions are two sides of the same coin. A gratuitous idea, like so many others, since the content of the standard model, described in the previous chapters, shows no such symmetry. Yet, the search for supersymmetry mobilized a disproportionate number of physicists worldwide for decades, with a significant expenditure of funding and resources, and in return, the diversion of grants, positions, and resources from all other sectors of high-energy physics research. Needless to say, the LHC never found the slightest trace of supersymmetry.

But all of this did not make headlines. The general public was not informed. And it was never mentioned that the Higgs boson had been predicted nearly fifty years earlier. Why? Because readers or viewers might have inquired: How did scientists spend their time during the intervening years? The answer is as follows: plunging headlong into studying a series of anticipated failures. Not only string theory, loop quantum gravity, and holography, but also supersymmetry and the alleged grand unification, i.e., the idea that the four fundamental interactions of nature should somehow be unified, interrelated by a symmetry. Needless to say, no sign of grand unification has ever been revealed.

The pursuit of symmetries that do not exist has blinded scientists for decades. Many human beings would like nature to be more symmetrical than it actually is (why is that? one wonders), but nature has never shown sensitivity to human preferences.

And that's not all. My experience has led me to conclude that beyond the discovery of quantum gravity, it is not so certain that the future of humanity is a scientific one. In fact, to be quite frank, there are increasingly evident signs that, not from today, but for about forty years, we have literally gone backward, and reached a standstill, plunging into a sort of new Middle Ages.

Often, high-energy physics provides an early glimpse of what will happen to society as a whole. And if progress has stopped for forty years in particle physics (out of about four hundred years of modern scientific history), this means that, at least in this domain, the scientific era is now over. And it also means that the scientific era was, indeed, a parenthesis, nothing more. For forty years now, we have been in a new era, a pseudo-scientific one.

Imagine an airplane flying through the sky, leaving a trail behind it. Imagine that at some point, something or someone brings down that plane. For a while, the trail remains visible, but after some time, even that disappears. Well, the airplane represents the most advanced physics. It has been the engine of progress in most of human knowledge, which is the trail. Now the airplane is no longer there. Perhaps it fell on its own, succumbing to the weight of the difficulties it faced. Maybe it was shot down by poor choices made over the past decades. But the trail remained, and it will stay there at least for a while. The trail symbolizes the downstream, comprising a variety of applications in different fields, which provide an excuse for believing that progress is not completely halted. However, it's fair to say that we no longer live in the time when that airplane graced the skies. The 21st century is not the 20th century. Many so-called scientific discoveries are now good only for making journalistic scoops, even when they have no relevance or content, or are mere hypothetical speculations.

And even if we have clarified what quantum gravity is, that alone is not enough to get the plane back in the air. The practical difficulties of conducting experiments in high-energy physics are such that after taking fifty years to experimentally confirm the prediction of the Higgs boson, we might take five hundred years to confirm quantum gravity. And perhaps another five thousand for the next discovery. There is no guarantee that the distribution of discoveries will be regular over time. Nothing can assure us that there is a law of proportionality between the number of discoveries, weighted by their relevance, and the number of people engaged in research, or the time, effort, and funding devoted to that research. Nothing guarantees that the more we invest, the more we discover. The relationship between the number of discoveries and the variables at play could be highly penalizing for us. Given the numbers we face when venturing into the infinitesimally small, given that the transition from quantum mechanics to quantum gravity is measured by a factor of a billion billion, we cannot rest on our laurels, or take for granted that the future will always be interesting. One of the most boring futures ever could lie ahead of us, one that could last a very long time, or even indefinitely. These durations could be unbearable, even deadly for us. If we fail to leave our planet to colonize the universe, it's not certain that we'll have much time left. And even if the human species were willing to settle for a dull future, there is no guarantee that this future will be long. Signs of regressive trends are becoming increasingly apparent.

We have now occupied the entire planet, while space colonization is not progressing much. We risk remaining isolated here for a long time. It is possible that the human species, after developing a prosperous civilization, will experience a period of involution. In physics, this has already begun, as we have mentioned. The problem is that, in general, involution is much faster than evolution. It takes by surprise, leaving no way out. Once this path is taken, there is no guarantee of turning back.

Even worse, we might not even realize that we are going through an involution. In that case, there is little we can do to stop it, or change course. It's a bit like that famous frog placed in water, slowly heated, which only realizes the danger when it's too late, when the point of no return has been crossed and there is nothing more to be done.

We cannot exclude the possibility that our future witnesses a regression of civilization to a primitive state, without us even realizing it. Wrongly, we believe that in our time, it is impossible to lose the knowledge accumulated over centuries of progress, given the technological means at our disposal, with which we can store, copy, and disseminate documents without limits. But the trap lies where we least expect it. Indeed, the loss of knowledge can sneak in and spread in various forms, including the most aggressive one, which is the overproduction of pseudo-knowledge, especially today.

It will be increasingly difficult to discern real knowledge from useless production. For example, today, in high-energy physics, many more scientific papers are produced than before, but progress is minimal, if not absent. The few articles written decades ago allowed for much more progress than the many articles published today. Nowadays, the numerous irrelevant articles form an ocean that drowns out the few relevant ones, if there are any. Even if discoveries are made, it is very difficult to become aware of them. And the situation is rapidly worsening.

This is an involutionary regression. Knowledge is buried beneath an ocean of irrelevant discussions, much like in the Middle Ages. And since we don't even realize what is going on, we can't do anything about it. Instead, we spend our time inventing objective criteria to evaluate production and research, such as counting the number of citations received by a scientific paper. Well, according to these criteria, one could say that we are going through one of the most culturally flourishing and productive periods in human history! But the reality is exactly the opposite: these criteria are fallacious.

For example, after two or three generations dominated by string theorists, the new recruits of theoretical physicists now have a very superficial knowledge of quantum field theory. They are not even capable of appreciating a potential discovery in such a research area. This says a lot about the possibility of gradually losing knowledge without even realizing it, even if the knowledge is actually not destroyed, but still present, in front of our eyes and at our disposal. Nonetheless, it is inaccessible, because we can no longer see it.

Many ancient texts remained buried in the most remote corners of monasteries throughout the Middle Ages. Scientific knowledge will meet a similar fate, and be inevitably forgotten. Perhaps it will remain buried for millennia in a USB drive, in a computer's memory, in an obscure corner of the web, or worse, the dark web, with slim hope of a future awakening of civilization. The overproduction of pseudo-knowledge will make it virtually impossible to discern real knowledge for who knows how long.

Yes, every form of life, every species, every individual, every choice is an attempt, a roll of the dice. As such, its presence in the universe has a beginning, a course, and an end. At some point, it runs out. Life, as a titanic endeavor to amplify quantum indeterminacy from microscopic to macroscopic scales, is fragile and unstable, constantly susceptible to the sudden transition from the expansion phase to the regression or even disintegration phase. This applies to the individual, but also to civilization, to the species, to life on a planet. For this reason, before departing, we should try to accomplish our mission, which is to create new forms of life, as different as possible from our own.

ABOUT THE AUTHOR

Damiano Anselmi is a theoretical physicist specializing in quantum field theory and quantum gravity. After earning his Ph.D. from ISAS Trieste, he conducted research in various parts of the world, including Harvard, École Polytechnique, CERN, the Chinese Academy of Sciences, Perimeter Institute, and the University of Pisa. He has authored numerous scientific publications in top international specialized journals.

To follow his research, you can visit the website http://renormalization.com, or check out his YouTube channel @QuantumGravityChannel.

www.ingramcontent.com/pod-product-compliance
Lightning Source LLC
Chambersburg PA
CBHW021958170526
45157CB00003B/1050